Housing and Asthma

The UK has by far the highest prevalence of asthma in the world. Over the past 25 years the incidence of asthma episodes has increased by a factor of three to four in adults and six in children. Why should this be? Although allergic disease is on the increase across the developed world, what factors are unique to the UK that can be identified as key causal mechanisms? This book reviews the evidence which has emerged from environmental science, respiratory medicine, immunology and building design, and carefully assembles the pieces of the jigsaw. A clear picture emerges.

The rise of the asthma epidemic can be traced back to the oil crisis in the mid-seventies which produced a drive for energy efficiency. Increasing levels of insulation, combined with double-glazing and the sealing of open chimneys, dramatically reduced domestic ventilation rates, which in turn produced warm, humid and polluted indoor environments. Such conditions proved ideal for the colonisation and proliferation of the house dust mite. This species excretes a range of allergens which have been identified, both as a causal mechanism in the aetiology of the disease and as irritants likely to trigger and exacerbate asthmatic symptoms. Eight out of ten asthmatic children in the UK are allergic to these proteins.

The book reports on the positive health benefits achieved in an interventionist trial that remediated a range of dwellings and successfully reduced HDM allergen levels, while improving indoor air quality. If the negative impacts of poor indoor air quality on respiratory health are to be avoided, the adoption of a preventative strategy will require a new approach to house design, construction and specification. The book contains details of how to build energy-efficient dwellings while maintaining healthy indoor air quality and demonstrates the fiscal, environmental and health benefits that can accrue. Such an approach must be adopted by those involved in the production of the built environment, if we wish to save the next generation from a lifetime spent wheezing.

Stirling Howieson is a Chartered Architect and Director of the Centre for Environmental Design and Research (CEDAR) at the University of Strathclyde.

Housing and Asthma

Stirling Howieson

Spon Press
Taylor & Francis Group

LONDON AND NEW YORK

First published 2005
by Spon Press
2 Park Square, Milton Park, Abingdon, Oxon OX14 4RN

Simultaneously published in the USA and Canada
by Spon Press
270 Madison Avenue, New York, NY 10016

Spon Press is an imprint of the Taylor & Francis Group

© 2005 Stirling Howieson

Typeset in Sabon by
Integra Software Services Pvt. Ltd, Pondicherry, India
Printed and bound in Great Britain by
TJ International Ltd, Padstow, Cornwall

British Library Cataloguing in Publication Data
A catalogue record for this book is available from the British Library

Library of Congress Cataloging in Publication Data
Howieson, Stirling.
   Housing and asthma / Stirling Howieson.
      p.   cm.
   Includes bibliographical references and index.
   ISBN 0–415–33645–7 (hb : alk. paper) —
   ISBN 0–415–33646–5 (pb : alk. paper)
   1. Asthma—Great Britain.   2. House dust mites.   I. Title.
   RA645.A83H69 2005
   614.5′9238′0941—dc22

                                                        2004012514

ISBN 0–415–33646–5

All the way to Munich, Vinnie guides the steering wheel with his elbows so that he can tap with his drumsticks on every hard surface. He gasps bits of songs, Mister Whatyoucallit whatcha doin' tonight, and bap bap da do bap do do de do bap, to go along with the beat and then he's so excited the asthma hits him and he's gasping so hard he has to stop the jeep and pump his inhaler. He rests his forehead on the steering wheel and when he looks up there are tears on his cheeks from the strain of trying to breathe. He tells me I should be grateful all I have is sore eyes. He wishes he had sore eyes instead of asthma. He could still play the drums without stopping to take his goddam inhaler. 'People don't appreciate not having asthma. They sit around moaning and bitching about life and all the time breathing, breathing nice and normal and taking it for granted. Give 'em one day of asthma and they'll spend the rest of their lives thanking God with every breath they take, just one day.'

Contents

Acknowledgements

I am indebted to a large number of individuals for their support and counsel. To Geoffrey Brundrett, Past President of the Chartered Institution of Building Services Engineers, for setting me off in this bizzare direction and for his continuing acerbic and quick-witted analysis.

To the members of the research team, Prof George Morris of the Scottish Centre for Infection and Environmental Health, Dr Charles McSharry, Clinical Immunologist at the Western Infirmary and Dr Eddie McKenzie, Statistician, University of Strathclyde, I offer my thanks for keeping the research vehicle on the right road. I may have been driving the bus, but there were many dead ends that appeared to be attractive routes. The value of such a multi-disciplinary approach cannot be overstated.

Most of all I am indebted to my research assistant Dr Alan Lawson who has worked alongside me for six years, mining a vast quarry of literature and sorting the wheat from the chaff. This was undertaken while vacuuming and steam cleaning hundreds of carpets in our relentless and merciless pursuit of the dreaded house dust mite.

There have been many contributors to the funding pot. Energy Action Scotland, a charity campaigning to end fuel poverty, provided an initial contribution of £45,000, which levered in funds from EAGA Charitable Trust, Scottish Power, Transco, Lanarkshire Health Board, North and South Lanarkshire Councils and Baxi Clean Air Systems Limited. Strathclyde University has also made significant research and development funds available. The second research phase now in progress is also supported by the Chief Scientist's office at the Scottish Executive. To date, close to £1 million has been contributed from all sources.

Finally, to the patients who volunteered for what must have appeared initially, as a 'snake oil salesman' approach to healthcare, I offer my thanks. It was most rewarding to see the quality of their lives improving with their increasing lung function. Breathing is an undervalued activity and should not be taken for granted.

Dr Stirling Howieson
BArch(Hons) DipArch MPhil PhD C.Eng ARIAS FCIBSE ILT
Chartered Architect and Chartered Engineer

Introduction

In February of 2004 the Global Initiative on Asthma[1] (GINA) reported that 18.4% of Scots suffer from asthma. This compares with 15.3% in England, 10.9% in the US, 6.9% in Germany, 6% in Belgium, 4.5% in Italy, 2.3% in Switzerland and 0.7% in Macau. The report also claims that 35% of Scottish 13–14 year olds, experienced 'wheezing' over a 12-month period. Of the 84 countries surveyed, Scotland had the highest prevalence of symptoms in this group and it appears from the initial data reporting symptoms in 6–7 years olds (which were 5% higher) that the disease remains on the increase. This summary report supports the findings of a previous study into asthma prevalence in 57 countries. The ISAAC[2] study, published in 1998, confirmed that the UK had the highest prevalence of asthma symptoms in 13–14 year olds in the world. Over the last 25 years the incidence of asthma episodes is three to four times higher in adults and six times higher in children. Although the disease is on the increase in many countries, Scotland is at the top of the scale. Why should this be? What features are specific to Scotland that can be identified as key drivers of the current asthma pandemic and, more importantly, what can be done to either slow the increase or reverse this trend?

The built environment and the construction industry have a poor record in research and development and within this limited body of research, housing conditions – particularly with regard to internal air quality – do not appear to have been a priority. Yet this is where most of us will spend the majority of our lives. The Institute of Medicine in the United States recently published an assessment of asthma risk and indoor air quality entitled 'Clearing the Air'. This systematic literature review identified only one substance where there was sufficient evidence to implicate it, both as a causal mechanism in the aetiology of the disease, and as an irritant likely to trigger and exacerbate symptoms. The substance was house dust mite (HDM) allergens. HDMs eat human skin and like most arthropods, thrive in warm humid conditions. As our dwellings have become warmer, due to increasing insulation standards and energy efficiency, they have also become much 'tighter' due mainly to the adoption of certain construction techniques, double-glazing and the sealing or exclusion of open chimneys. This is likely to have resulted in much lower ventilation rates, which in turn impact on indoor air quality. The task presented was to measure and assess the outcomes of these changes on respiratory health.

In 1990 the author was the project architect for a major refurbishment of 42 dwellings in the east-end of Glasgow, which formed part of a European Solar Demonstration Project. After a period of decanting, many residents on re-occupation, reported that their asthma symptoms had reduced in severity. This provided the spark for this research programme. The hunt was on to identify the mechanisms at work, and replicate this apparent miracle cure. The three-storey tenemental, flatted blocks, which were the subject of remediation, were hard to heat and suffered from a high incidence of dampness. The improvement and repair measures included insulating the external envelope, installing central heating and forming glazed utility balconies. Mechanical extract fans removed humid air from the bathrooms and kitchens, creating a slightly negative pressure in the dwelling and prewarmed air was drawn from the solar heated glazed balconies and common stairwell.

The refurbishment fundamentally changed the hygro-thermal conditions within the dwellings. The average internal whole-house temperature was raised to achieve comfort conditions across the dwelling, and moisture from cooking, clothes drying and bathing was – in the main – extracted at source. The dwellings were no longer cold and damp. They were now warm with relatively low humidity. In addition to this many residents had taken the opportunity, to renew carpets, beds and soft furnishings – all items previously identified as suitable for dust mite colonisation and proliferation. The major allergen reservoirs had thus been eradicated in many instances, while the new central heating and ventilation regime, ensured that the relative humidity seldom reached the critical equilibrium required for the HDM. A programme that was primarily concerned with energy efficiency had thus served to build a hypothesis and the research question: Are our homes causing the asthma pandemic?

Although the HDM appears to be a prime candidate implicated as a causal factor, there are a number of questions that have to be considered. This requires a review of housing legislation, economics, design theory and constraints, material and component specifications, changing living patterns and social norms. It requires six key research questions to be addressed:

1   Do HDM allergens cause asthma?
2   What is the likely scale of human exposure to these allergens in Scotland?
3   What historical changes have lead to the HDM species colonising and proliferating in the domestic environment?

4   What can be done to reduce allergen levels in our homes?
5   What other factors are implicated in indoor air quality and respiratory health that may influence the efficacy of the remediation programmes and measured outcomes?
6   What are the cost benefits of such a preventative approach?

This work initially reviews the aetiology of asthma and allergy, and looks at the possible respective roles played by genetics, diet and the environment. Chapters 2 and 3 concentrate on the ecology and physiology of the HDM and its allergens, while Chapter 4 reviews the evidence base investigating why our dwellings appear to be providing ideal environments for the HDM. It is in essence, a potted history of Scottish urban housing, as it has developed in the late nineteenth and twentieth centuries. It highlights the main legislative drivers and how the failure of free-market economic models required massive state intervention in the sector. The pressure to reduce internal volumes and resist raising standards is an inherent part of *laissez faire* economics. It also has a focus on how these factors have driven major generic changes in housing design, specification and use patterns that have had a direct impact on respiratory health.

Chapter 5 looks at the likely scale of the changes that have occurred in the housing stock during the twentieth century, and the impact these changes are likely to have had on ventilation and vapour diffusion rates. Five common generic house types are modelled using computer-based dynamic simulation techniques to assess the likely scale of changes in domestic ventilation regimes. Although such an approach must be heavily qualified, it does allow the scale of the changes to be norm referenced.

This leads directly to the testing of the primary hypothesis: that our dwellings have become warmer and more humid – conditions under which dust mite colonies can thrive. Chapter 6 details a research protocol to test the efficacy of a remediation programme designed to eradicate HDM allergens and suppress internal relative humidity.

Subsequent to reporting and discussing the results of this interventionist trial, several chapters are dedicated to identifying the confounding variables. This literature review has a focus on more recent papers, discussing likely airborne determinants such as indoor gases, microbes, moulds, particulates, endotoxins and volatile organic compounds (VOCs). It is primarily an attempt to evaluate the most influential factors and assess which ones should be scoped to provide secondary outcomes and generate further hypothesis for future research. This secondary literature review underpins a revised research protocol for a second experimental phase that went on-site in February 2004 and will report December 2005.

Chapter 10 concentrates on ventilation rates and reviews the major UK studies that have been undertaken in this area. It is clear that the background ventilation characteristics of the domestic stock remain under-researched. With reference to the simulation work reported in Chapter 5, it appears that over-ventilation may reduce a dwelling's energy efficiency and comfort, but it is unlikely that such 'leaky' dwellings will suffer from 'poor' air quality and dust mite infestation. The challenge for building designers is thus to address both problems and produce energy-efficient dwellings with 'healthy' air quality.

The outcomes of the initial trial have provided a sufficient level of evidence to support the development of a new dwelling prototype. This design reverses the principles of timber frame construction, encapsulating 'heavy-weight' mass within a thick layer of insulation. Such thermal capacitance reduces diurnal fluctuations and will ensure a more even internal temperature distribution. This in turn suppresses relative humidity below the critical equilibrium threshold required for dust mite colonisation. By maintaining the ventilation rates at $c.$ 1 ach^{-1} other indoor pollutants implicated in the aetiology of asthma – or the exacerbation of the symptomatic state – will also be expelled or diluted. This work seeks to report the development of a key knowledge base. It draws on insights gained in previous research activities and practice and sets the context for future activity.

The USA Institute of Medicine[3] were unequivocal in their conclusions as to the best way forward,

> The committee believes that better communication between medical, public health, behavioral science, engineering and building professionals is likely to result in more informed studies on the causes of asthma and the means to limit problematic exposures. Increased research efforts are needed to address the characteristics of healthy indoor environments. Asthma research clearly needs inter-disciplinary involvement – not only of clinicians, immunologists and researchers in related biologic areas – but also engineers, architects, materials manufacturers and others responsible for the design and function of indoor environments.

Architects and engineers will always tend to be generalists. A specialist knows a lot about a little. A generalist knows a little about a lot. The task has been akin to assembling the pieces of a jigsaw. First you need to know where to look and have some idea of how to edit and filter the pieces, focussing on those that appear most relevant. You then have to assess what pieces are missing and try to fabricate something that will fill any large voids. Although the final assembly may be less than a perfect fit, when you stand back from this collage, the bigger picture becomes obvious and undeniable. There is a growing and compelling body of evidence that supports – at least on the balance of probability – the hypothesis that our homes are the most influential, single identifiable factor, driving the current asthma pandemic in the UK. This book is no more than a snapshot of work in progress. There remains much to be done.

References

[1] Masoli, M., Fabian, D., Holt, S. and Beasley, R. Global Burden of Asthma, Global Initiative for Asthma, Medical Research Institute of New Zealand and University of Southampton, 1994, Summary report (www.ginasthma.com).

[2] The International Study of Asthma and Allergies in Children (ISAAC) Steering Committee, Worldwide variation in prevalence of symptoms of asthma, allergic rhinoconjunctivitis, and atopic eczema: ISAAC, *The Lancet*, 25 April 1998, London, pp. 1225–1232.

[3] Institute of Medicine, Clearing the Air, Committee on the Assessment of Asthma and Indoor Air, National Academic Press, Washington, 2002, ISBN 0–309–06496–1.

Chapter 1

The aetiology of allergy and asthma

The UK has the highest prevalence of asthma symptoms in 13–14 year olds in the world (Figure 1.1). Over the last 25 years the incidence of asthma episodes has increased by a factor of three to four in adults, and six in children. Although the disease is on the increase in many countries, the UK has by far the highest prevalence. Why should this be? What features are specific to Britain that can be identified and isolated as key drivers and, more importantly, can anything be done to either slow the increase or reverse the trend?

There are a variety of non-environmental factors that have been implicated in predisposing individuals to atopic reactions. It is thus necessary to discuss the role they may play and their significance in the aetiology of the disease, before turning to the specifics of the domestic environment. A range of hypothesis have been postulated to account for the dramatic increase in allergic disease: increased exposure to perennial allergens; housing changes;

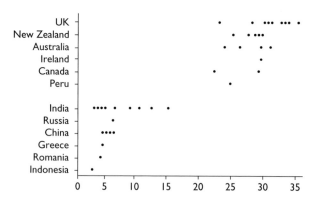

Figure 1.1 Prevalence (%) of asthma symptoms in 56 countries (13–14-year-old children n = 463,801 – extract highest and lowest six).
Source: ISAAC Steering Committee.[1]

more time spent indoors resulting in increased exposure; decrease in bacterial and other infections; improved hygiene; immunisation; prophylactic use of antibiotics; change in gut flora caused by antibiotics and/or diet; decline in physical exercise; high fat/salt diet and the loss of a specific protective effect.

What is allergy?

Allergy is a form of immunity in which the natural defence mechanisms of the body have misfired, and instead of protecting, actually cause disease. The word allergy itself means an 'altered capacity to react'. The term has been expanded by Davies and Ollier,[2] "Allergy is the inappropriate and harmful response of the immune system to normally harmless substances."

The immune system is effectively the department of defence of the complex chemical plant that comprises the human body. It guards the body from foreign invaders that enter through the skin or respiratory system by mobilising cells designed to attack and kill any such intruder. It is the over-reaction of the immune system to relatively harmless foreign substances, by producing too many of these cells, that causes allergic reactions. Allergy has four distinctive features:

1   Allergy only develops after the substance causing it has been encountered.
2   Only a proportion of those exposed to the substance develop the disease.
3   It involves specific proteins (antibodies) present in the body.
4   Chemicals (histamine) are released from mast cells.

What is asthma?

Asthma comes from the Greek term meaning 'hard breathing'. The hallmarks of asthma are wheezing, cough and breathlessness. The problems in diagnosing the condition are twofold; each of the above symptoms can occur separately or together in other chest conditions. Second, there is no firm marker to identify when the severity of symptoms becomes strong enough to justify labelling the sufferer with asthma. This can be further confused by the variability of symptoms, characteristic of asthma. Commonly afflicting the young, it can appear in later life where the condition tends to persist into old age. Children's asthma can be particularly capricious and both bacterial and viral infections appear to play a significant role in their symptomatic state.

The respiratory system

Davies and Ollier[2] describe the respiratory system as a process designed to filter and moisten the air entering the lungs and transfer gasses to and from the bloodstream. To function properly, all cells in the body require oxygen. In combination with sugars, oxygen releases energy, with the production of carbon dioxide as the waste product – a process called 'aerobic respiration'. The purpose of breathing is to extract the oxygen from the incoming air and exhale carbon dioxide. On breathing in about 300 million air sacs (alveoli) bud, producing an external appearance very much like bunches of grapes. The internal surface area of these air sacs if spread out would cover an area about the size of a tennis court. The blood vessels supplying the lungs branch into a myriad of tiny-walled capillaries which course over the surface of the alveoli, through which the blood in the capillaries picks up oxygen and gives off carbon dioxide before returning replenished to the heart.

Asthma is caused by narrowing or blockages in the airways supplying the lungs, which means that breathing is constrained wheezing is the result of air being forced through a narrow airway. The narrowing or blockage of the airways can be due to several processes:

- excess mucous in the bronchi;
- the swelling of internal layers of the bronchi;
- contraction of muscles lying in the walls of the bronchi.

The bronchi are made of several layers and the bronchial mucosa is a very thin layer of cells pointing inwards, lined by extremely fine hairs (cilia), which beat in rapid motion upwards towards the mouth, carrying with them a fine layer of mucous. Clusters of mucous glands within the lung layer produce this mucous, which traps any particles inhaled into the lungs and carries them – due to the wafting action of the cilia – away from the deeper parts of the lungs. Excess mucous production due to inflammation of airway linings can cause lung passages to narrow, leading to breathing difficulties and wheezing.

To become allergic the person must have the propensity to produce excess antibodies. This is known as allergic predisposition or atopy. A patient can present with a type I over-reaction to an allergen if they have already manu-factured the Immunoglobulin E (IgE – proteins that function as antibodies) antibody to that allergen. Symptoms occur within seconds or minutes of exposure and are caused by the binding of mast cells to the IgE antibodies and the invading allergen. This action produces strong chemicals such as histamine, whose function is to neutralise the invader. The effects of the release of histamine and other such chemicals are:

- The area exposed becomes inflamed due to increased blood flow.
- Fluid leaks from blood capillaries into the surrounding tissue and causes swelling.

- The muscles around the air passages contract, resulting in narrowing of the bronchi in turn leading to asthma.
- The flow of mucous is increased which causes sneezing and coughing.

Histamine levels can typically increase by up to a factor of five during such incidents. These reactions are essential in fighting off infection but are exaggerated and inappropriate in allergic individuals, who are over-reacting to what is normally a harmless substance.

There are however some types of reaction (type IV) which can take hours or even days to manifest after exposure. Contact dermatitis and occupational asthma are conditions where this type of reaction occurs. Any type I reaction can be followed by a type IV reaction. In a type IV reaction, the mast cells release chemicals which summon disease-fighting white blood cells (eosinophils) to the reaction site. This results in further inflammation and additional damage to the protective linings. These eosinophils take some hours to reach the reaction site, hence the delay. This delay can confuse many patients who believe they are suffering from a cold or infection and do not associate the symptoms with a past exposure. Symptoms such as sneezing, breathlessness, skin problems, blocked nose, coughing, asthma, itchy skin or feverishness could be the symptoms of an allergy of which the patient is completely unaware. In type IV reactions the cause and effect is less obvious and therefore are less often reported and/or diagnosed. Some of the increase in asthma may simply be due to the increased awareness of the condition. As there are a range of good drug treatments available the temptation for GPs may be to adopt the precautionary principle; prescribe first, diagnose later.

The respective role of genetics, diet and the environment

Marsh[3,4] maintains that the development of allergy is the result of the interplay between genetics (genes controlling total IgE antibody levels and genes determining a generalised immune hyper-response) and environmental factors such as childhood infection, diet, allergen exposure and even psychological factors. To investigate the relative importance of each of these factors, several trials have been conducted on twins. A study by Lubs[5] on 7000 pairs of same-sex twins could not distinguish between environmental factors and genetic factors, concluding that both were important in the development of asthma. Another study by Wutrich et al.[6] on fifty sets of twins, where at least one twin had a history of atopy (an atopic person is someone that is predisposed to allergy), indicated that although the tendency to IgE production is genetically governed, environmental exposures determine what specific antibodies are produced. It seems that the capacity to react is inherited, but exposure to antigens determines whether overt allergy results and to which substance the individual reacts.

Specific allergy therefore cannot be transferred from parents to their offspring, only the ability to react is inherited. Children of parents with allergies are more likely to have elevated antibody levels to common allergens. The risk factors for ceding allergic offspring were calculated by Davies and Ollier[2] to be: neither parent allergic – risk factor of 10–20%; one allergic parent – risk factor 30–50%; two allergic parents – risk factor of 40–75%.

Breast feeding

The hypothesis that breast feeding has a prophylactic effect against allergy does not appear to be completely resolved. The first and largest study in this field by Grulee and Sanford[7] followed 20,000 infants from birth to age nine months. It found that compared to breast-fed babies the incidence of eczema was twice as high in partially breast-fed babies (soya milk supplement) and seven times as high in bottle-only-fed babies. The importance of these results is highlighted by the fact that it was conducted at a time when any such relationship was not suspected and therefore parental bias in the study was unlikely.

It has been known since the 1920s that food antigens can cross the breast milk and subsequently sensitise the nursing infant.[8–11] In a more recent study, Cogswell and Alexander[12] found that 17 out of 19 breast-fed infants showed positive skin tests to eggs, as opposed to 1 out of 13 of those bottle-fed. Businco, Marchetti and Pellegrini[13] took the underlying hypothesis one stage further by restricting infants to either breast feeding or bottled milk and limiting the mothers to 200 ml of milk per day and two eggs per week. Additionally, in the study group infants, no weanings were introduced for the first six months. The study confirmed a higher rate of atopy in bottle-fed babies as opposed to breast-fed (6/36 pure breast-fed, 6/25 soya milk and breast milk, 15/41 cows' milk). In a later trial[14] he used a similar protocol but included the following environmental measures: no smoking in the house; no pets; control of dust (unclear whether HDM) and no day care attendance before three years. Out of the 100 breast-fed infants in the study, 18 developed atopic symptoms. This compares with 8 out of 79 partially breast-fed and 25 out of 65 fed on bottled formula milk. Additionally the study demonstrated that males develop atopic disease more frequently than females.

Other studies[15,16] have confirmed that dietary and environmental measures were able to prevent the onset of atopy in high risk babies, at least until the age of three years and eight months. They also claimed that the time of starting and duration of the protective measures are important, and only infants breast-fed for a long period of time and in whom the re-introduction of solid foods is delayed can produce significant protection. Aeroallergen avoidance and the mother's diet were also shown to be important variables influencing the odds of the child developing allergy.[17–19]

Exposure to outdoor allergens

Croner and Kjellman[20] correlated cord blood IgE with month of birth and reported that environmental factors are specific, and family planning is important for those with a high genetic risk of atopy. As the severity and incidence of atopy rises – therefore increasing the number of future sufferers – this may become increasingly important. Atopic disease seems to develop more often in children born in the months preceding the peak pollen season[21–25] and there is a clear implication that the month of birth influences the incidence of allergen exposure. Quoix et al.[26] claimed that this was only true for patients with mould sensitivity, however, further studies have found this to be the case for pollen related allergies, although variations in the seasonal peak can affect results.[27,28]

On the night of 24–25 June 1994 a thunderstorm over southern England triggered a tenfold increase in acute asthma attacks presenting at accident and emergency departments.[29] It was hypothesised that osmotic rupture of pollen grains during the thunderstorm released submicroscopic starch granules which could penetrate much deeper into the lungs causing an immediate allergic reaction in atopic individuals.

Viral infections and the hygiene hypothesis

Busse[30] in a review of related studies, concluded that the evidence pointed to a common pathway for the development of airway hyper-reactivity during respiratory viral infections. Infections can cause inflammation of the airways, in turn producing increased bronchial reactivity and obstruction. In contrast, a study by Martinez[31] suggested, from an immunological standpoint, that frequent infection in early infancy would lead to unfavourable circumstances for allergic sensitisation. Shaheen[32] produced evidence to suggest that measles infection and possibly other respiratory viruses contracted in the first year of life, may protect against the onset of asthma. Infections appear to make the immune system more robust and less likely to react to what are, in effect, harmless substances. We have a powerful immune system that has evolved to fight regular bacterial invasions. If we are suddenly removed from these 'dirty' environments the immune system may not simply 'switch off', but instead hunt for other invaders that it has previously considered to be minor and unimportant to target.

Other factors, such as the over-use of antibiotics, which Holt and Sly[33] has claimed alters intestinal bacterial flora, and the strong inverse relationship, shown by Shirakawa et al.,[34] between the magnitude of a skin BCG response in Japanese children and the later development of asthma are persuasive arguments in favour of the 'hygiene hypothesis': that the increasing prevalence of asthma is due to decreasing prevalence of infections and exposure to bacteria in early life.

There does not appear to be a critical mass of evidence to substantiate a causal relationship between any of the above features and the development of allergy. Holgate,[35] in a meta-analysis of related studies, concluded that a clearer understanding of the cellular and molecular mechanisms involved in disease induction and progression is needed to fulfil the promise of treatments that will modify the disease outcome, rather than simply suppress the inflammatory process.

The healthy pet-keeping effect

The role of the indoor environment in asthma is of major concern because a large proportion of children and young adults with asthma are allergic to allergens found indoors. Some recent studies have claimed that children raised in a home with animals are less likely to become allergic. It has been claimed[36–38] for some years that airborne exposure to cat allergens in a house was quantitatively much higher than the exposure to dust mite allergens, however, the prevalence of sensitisation to cats is generally less than sensitisation to mites. This raises questions about the mechanisms by which children and adults 'tolerate' living in a house with a cat, without becoming allergic, even when they are from genetically atopic families and are allergic to a range of other substances. The evidence tends to suggest that the specific type of allergen present is the most important factor determining health outcomes.

The 'healthy pet-keeping effect' has recently been challenged by Bornehag *et al.*[39] who claimed that such cross-sectional or cohort studies cannot be used to investigate associations between pet-keeping and allergic disease, as atopic individuals will naturally avoid or get rid of pets. The study which looked at 10 851 children in Denmark concluded that those who choose to keep pets are those that are not susceptible to allergic reactions. Children that had pets at the time of birth or later – but not during the study period – had the highest prevalence of symptoms. No healthy pet-keeping effect therefore exists. It is simply the statistical outcome of cohort selection bias.

There have been major changes over the last 40 years that could have influenced lung pathology but are not related to the indoor environment. Examples include: the introduction of broad spectrum antibiotics in 1960, widespread use of immunisation and extensive use of medicines such as paracetamol (it is claimed that paracetamol breaks down into oxidising free radicals which are then able to cause cell damage and increase the burden and sensitivity to particulate matter $PM_{2.5}$) and/or the inhalation of antihistamines.

Geographic-specific factors

Similarly, it has been argued that major changes in cleanliness have allowed the increase in allergic disease as the body's powerful immune system is under-exercised and thus it pursues harmless invaders. One version of the cleanliness hypothesis argues that exposure to mycobacteria or other infectious agents can bias the whole immune system towards an allergic response. This argument implies that many or most non-allergic individuals also respond to common allergens such as dust mite. Platts-Mills et al.[40] claim that increased 'cleanliness' may well explain the rise in allergic rhinitis between 1873 and 1935, but seems unlikely to explain the increase in asthma between 1960 and 2000. Most of the effects that could explain a change of this scale did not occur consistently over the period when asthma increased, i.e. 1960–2000. Platts-Mills et al. considered three examples of communities where asthma has increased: the suburbs of Sydney where there is a high prevalence of childhood asthma; those areas of Scandinavia where dust mites are not a significant allergen; and the American inner cities where children living in poverty suffered major increases in prevalence, severity and mortality from asthma over the period 1965–1995. These examples represent dramatically different climates, housing conditions, dietary changes, and in keeping with that, have different patterns of allergen sensitisation. In each of these areas asthma among children over three years of age is strongly associated with skin sensitivity or IgE production to relevant indoor allergens. In Sydney, similar to the UK, asthma correlates very strongly with IgE or skin tests to dust mite. Indeed, Peat and Li[41] calculated that in areas of high exposure to mite allergens the odds ratios for asthma among mite allergic individuals were greater than ten. In these areas there is a very high prevalence of cats in houses, but mite sensitisation is consistently more important than cat sensitisation. As previously discussed, there is a clear implication that mite allergen is more potent and that any tolerance effects that have been claimed for cat exposure cannot overcome sensitisation to HDM proteins.

In Scandinavia asthma has increased over the same timescale but the actual numbers are much lower. In addition, it is possible that severity is less due to inhaled steroid use having effectively controlled hospital admissions. In Northern areas of Scandinavia there are very few mites or cockroaches in homes and sensitisation to the allergens of domestic animals is the dominant association with asthma. Children raised in homes with a cat have consistently shown decreased sensitisation to cat allergens. In a cohort study in Norbotten by Ronmark et al.,[42] sensitisation to cat allergens was the strongest risk factor for incident physician-diagnosed asthma (odds ratio of 7.4). Living in a house with a cat was shown by this study to be a strong negative predictor of incident asthma. Perzanowski et al.[43] also claimed this effect exists among children in the United States with a family history of asthma.

It appears that exposure to proteins derived from cats does not produce as severe asthmatic symptoms as would be predicted from allergen exposure measurements but asthma has increased in countries where cat allergens are the dominant sensitisers.

American cities have experienced a major increase in asthma as judged by prevalence, hospitalisation and mortality figures.[44,45] In each case the numbers are consistently worse among African Americans living in poverty. Although there are many possible explanations for this, increased cleanliness seems the least likely. The major allergens in the cities vary. In New York both cockroach and rodents have been implicated. By contrast in Atlanta,[46] dust mite and cockroach allergens dominate sensitisation. In Chicago[47] a wide range of allergens seem to be relevant to asthma but those derived from the German cockroach are most important. From these and other studies[48,49] it appears likely that cockroach allergens are very potent sensitisers and that high exposure is an important cause of severity.

In each of these three examples there appears to be a strong case that asthma has increased and there are well-defined sources of indoor allergens associated with asthma. The results do not explain why asthma should have increased over such a consistent timescale in each environment. Many other changes may have contributed: indoor pollutants, dietary factors, immunisation practices, antibiotic use and outdoor pollutants. The changes in housing or dust in Scandinavia cannot be compared with changes in South Carolina. Platts-Mills[40] has claimed that the big improvements in hygiene occurred in New York and Chicago at least 40 years before the current increase in asthma. In the US the nematode parasites and malaria were eradicated by 1920; water supplies were clean by 1900 and chlorinated by 1930; separation of farm animals from the urban population occurred before 1900. New results[50] showing that high exposure to cat allergens can induce 'tolerance' among 15–20% of the population strongly argues against progressive increases in exposure as the explanation for the increase in asthma in Scandinavia.

Major questions remain, but central to these are the changes in lifestyle that have been so severe in the USA, but have also occurred throughout the Western world. These include sedentary lifestyle, changes in diet or excess food consumption and the closely linked rise in obesity. It may be that the significance of allergens and pollutants may be affected by many other lifestyle factors.

Prevalence of atopic disease

The past century has seen a dramatic rise in the prevalence and severity of atopic allergy in Europe and the USA.[51–53] At present around 40% of the human population has a hereditary predisposition to atopy.[51] Asthma is by far the most prevalent allergic reaction to the extent that Mygind[54] has produced evidence to support the view that it is now the most common chronic disease in children in the Western world. In the UK, the National Asthma Campaign[54] has claimed that by 1996, asthma had been diagnosed in 7.8% of adults and 12.5–15.5% of children in England. Similar findings were found in all age groups in Scotland with the exception of children, who had a slightly lower level. The severity of asthma has also increased as recognised in hospital admissions in England, Wales and the USA.[55] Death rates for asthma peaked in the 1960s and 1980s but have recently levelled off. Around 1500 people still die from asthma each year in the UK and Jackson et al.[56] have suggested that child mortality is on the increase.

Allergic diseases are stimulated by a reaction of the immune system to a foreign substance. Exposures to such substances are most likely experienced in the home. Lynden van Nes[57] has shown that the distribution of arthropods varies across Europe and the USA, due to internal hygro-thermal conditions influenced by external climate (Table 1.1).

It is clear from this distribution that areas where relative humidity is higher, possibly due to a maritime influence, dust mites – and the associated increased likelihood of asthma – are greater. From this evidence, dust mites appear to be the single most-common allergic stimulus. The importance of cockroach allergy in North America is the only possible exception to this rule. Kitch et al.[58] found in a recent US study of 499 households in Boston that lower family incomes, less maternal education and race/ethnicity (black or Hispanic versus white) were associated with a lower risk of dust mite allergen levels but an increased risk of exposure to cockroach allergens. In Western Europe, the dust mite appears unrivalled in terms of the proportion of atopic individuals that demonstrate positive reactions to its extracts. A causal mechanism, a dose–response relationship and an exposure scale have been established and agreed. It appears that the only remaining task is to quantify allergen levels in a statistically significant subset of dwellings and correlate the results with patient sensitisation and symptom severity.

A major review by Strachan,[59] however, directly contradicts the view that environmental factors have played a significant role. The paper concluded that it is unlikely that trends in either outdoor or indoor air pollution have contributed substantially to the rise in asthma, citing as an example Australia, where widely varying climatic conditions give rise to regional variations in mite populations, yet asthma prevalence in children is similar across all

Table 1.1 Skin prick tests signifying allergen exposure incidence

Arthropod allergens	UK	The Netherlands	Scandinavia	Mediterranean	USA
House dust mites	80	60–90	2–27	15–49	24–78
Storage mites	30	65–70	40–45	10	12
Cockroaches	10	10	10	10	7–69

states. The paper does admit that intervention studies may offer improved evaluation and that those studies published before his review had proved disappointing in terms of their attempts to reduce allergen exposure in the domestic environment.

Testing for allergies

A number of tests have been developed to identify the specific substances that cause allergic reactions in individuals and the severity of allergic disease. Mygind *et al.*[51] summarised these into three main categories.

Skin prick testing

Usually carried out on the inside of the forearm, or on the back of an infant, these tests involve the delivery of minute amounts of allergen through the skin by application of a tiny prick with a needle. Several different pricks may be conducted on the testing area, with each applied allergen being referenced to the respective marks. The chemical reaction that occurs (minor allergic reaction) causes the blood vessels to leak fluid causing a raised blister-like weal. A positive reaction noticed as intense itching will occur within 1–2 min of application, followed by a weal, which reaches a maximum within 10–15 min and is surrounded by a wide patch of redness and itching. The level of redness and size of weal resulting from either method of skin testing are analysed by the physician. A zero reading is observed if the skin appears normal i.e. no sign of reaction is visible. The severity of any reaction that is visible is ranked on a scale of 1 (lowest)–4 (highest). If the reaction is strong then the patient will be advised to cut out or avoid the offending substance. The problem with skin tests is that they may be affected by external factors such as diet or medicine e.g. if a patient is taking antihistamines, which block the effect of histamine and fail to inform the person performing the test, a false result would be produced.

Blood tests

PRIST stands for 'Paper Radio-Immunosorbent Test' and is used to measure the total level of IgE in the blood i.e. the total amount of antibodies present capable of causing allergy. They are more accurate than skin tests as they are not affected by external factors. The test itself is complex and involves the use of radioactivity. Results of total IgE below a certain level (usually 100 international units per millilitre of blood) means that the patient is non-allergic. RAST stands for 'Radio-Allergosorbent Test' and is used to measure the amount of IgE antibodies present in blood that relate to individual allergens such as pollen and the HDM.

Breathing tests

Air will normally flow unrestricted through the lungs, but in sensitised individuals inflammation of the linings of the lung reduces airflow. The degree of limitation is measured by tests of lung function. Peak-flow metres are the simplest devices used to measure lung function and they are used to evaluate the maximum rate at which air can be expelled from the lungs. Factors affecting peak flow include height, age, sex and severity of asthma. The more severe the asthma, the lower the peak-flow rate. A vitalograph is a device that can measure both the total volume and amount of airflow limitation of the lungs. The patient blows forcibly into the vitalograph and then continues to blow until all air is expelled from the lungs. The Forced Vital Capacity (FVC) is the total volume of air expelled over the course of the test, whereas the Forced Expiratory Volume in one second (FEV1) is the volume of air expelled in the first second of the test – both are affected by the same factors that affect peak flow. Ratios of FEV1 to FVC normally lie in the region of 80% but in asthmatic subjects this reduces to below 70% as asthma reduces FEV1, but FVC generally remains constant. In very severe cases this ratio can fall to 25%.

Asthma is usually diagnosed by observing symptoms and conducting breathing tests. If FEV1 initially falls, then returns to normal after inhaling a bronchodilator aerosol (reversibility), asthma is diagnosed. When individuals display symptoms not typical to asthma e.g. cough and no wheeze or shortness of breath, and normal breathing test results, other tests are required to prove the diagnosis of asthma. These tests can also be used to evaluate the level of excessive irritability in the airways, called hyper-responsiveness, which is present to some degree in all asthmatics.

To measure hyper-responsiveness patients are asked to inhale aerosols of a series of solutions of histamine or methacholine, which are gradually increased in strength until a mild attack of breathlessness or wheezing is observed. In these tests, known as provocation or challenge tests, the weaker the solution that produces the symptoms, the more sensitive a patient and in general, the greater the need for treatment.

Chapter summary

This chapter has summarised the key features of allergic reactions and discussed a list of possible non-indoor air-related causal factors in the aetiology of the disease. The factors affecting the onset of allergy appear to be a combination of a genetic predisposition and a subsequent exposure to allergenic material. Protective elements inhibiting sensitisation remain unclear. Allergic parents, breast feeding and month of birth, have all been implicated as having an important role, but there are few definitive answers. To simplify the respective roles that certain compounds may play, Tables 1.2 and 1.3

Table 1.2 The association between indoor biological and chemical exposures and the development of asthma

Biological agents	Chemical agents
Sufficient Evidence of a causal relationship	
House dust mite	(no agents met this definition)
Sufficient evidence of an association	
(no agents met this definition)	ETS in pre-schoolchildren
Limited or suggestive evidence of an association	
Cockroach in pre-schoolchildren	(no agents met this definition)
Respiratory syncytial virus (RSV)	
Inadequate or insufficient evidence to determine whether or not an association exists	
Cat/dog/cow/horse/domestic birds	NO_2/Nitrogen oxides
Rodents/cockroaches/endotoxins/fungi	Pesticides/plasticisers/VOCs
Chlamydia pneumoniae/trachomatis	Formaldehyde/fragrances/ETS
Mycoplasma pneumoniae	
House plants and pollen	
Limited or suggestive evidence of no association	
Rhinovirus (adults)	(no agents met this definition)

Table 1.3 The association between indoor biological and chemical exposures and the exacerbation of asthma in sensitive individuals

Biological agents	Chemical agents
Sufficient evidence of a causal relationship	
House dust mite/cockroach/cat	ETS (in pre-schoolchildren)
Sufficient evidence of an association	
Dog	NO_2/Nitrogen oxides
Fungi or moulds	
Rhinovirus	
Limited or suggestive evidence of an association	
Domestic birds	ETS
Chlamydia pneumoniae	Formaldehyde/fragrances
Mycoplasma pneumoniae	
Respiratory syncytial virus	
Inadequate or insufficient evidence to determine whether or not an association exists	
Cow/horse/rodents	Pesticides/plasticisers/VOCs
Chlamydia trachomatis	
Endotoxins	
House plants	
Pollen exposure in indoor environments	
Insects other than cockroaches	
Limited or suggestive evidence of no association	
(no agents met this definition)	(no agents met this definition)

summarise the outcomes of the USA Institute of Medicine[60] which categorised a range of compounds implicated in either causing asthma or exacerbating the symptoms. They used a five-point scale of evidence: (i) sufficient evidence of a causal relationship; (ii) sufficient evidence of an association; (iii) limited or suggestive evidence of an association; (iv) inadequate or insufficient evidence to determine whether or not an association exists; and (v) limited or suggestive evidence of no association.

Once an individual has become predisposed to allergy, subsequent exposure to a variety of allergens can stimulate the onset of allergic disease and associated symptoms. Asthma has grown to become the most common manifestation of allergic disease. Other allergic diseases such as rhinitis and eczema have also increased in prevalence and severity over the course of the twentieth century, however asthma has seen the greatest increase. Indoor environmental factors are known to trigger allergy, and may be of great importance as both sensitisers and triggers. Reducing such exposure may play the key role in reducing the prevalence of the disease and the incidence of asthmatic symptoms.

References

[1] ISAAC Steering Committee, *The Lancet*, Vol. 351, 1998.

[2] Davies R and Ollier S. *Allergy: The Facts*. Oxford University Press, Oxford, ISBN 0-19-261439-8, 1989.

[3] Marsh DG. Allergens and the genetics of allergy. In: M Sela (ed.) *The Antigens*. III. Academic Press, New York, 1975, pp. 271–359.

[4] Marsh DG, Meyers DA and Bias WB. The epidemiology and genetics of atopic allergy. *New England Journal of Medicine*, 1981, Vol. 305, pp. 1551–1559.

[5] Lubs MLE. Allergy in 7000 twin pairs. *Acta Allergoli*, 1971, Vol. 26, pp. 249–285.

[6] Wutrich B, Baumann E, Fries RA and Schnyder UW. Total and specific IgE (RAST) in atopic twins. *Clinical Allergy*, 1981, Vol. 11, pp. 147–154.

[7] Grulee CG and Sanford HN. The influence of breast and artificial feeding on infantile eczema. *Journal of Pediatrics*, 1936, Vol. 9, pp. 223–225

[8] O'Keefe ES. The relation of food to infantile eczema. *Boston Medical Surgery Journal*, 1920, Vol. 183, p. 569.

[9] Shannon WR. Demonstration of food proteins in human breast milk by anaphylactic experiments on guinea pigs. *American Journal of Diseases in Children*, 1921, Vol. 22, p. 223.

[10] Shannon WR. Eczema in breast-fed infants as a result of sensitisation to foods in the mothers diet. *American Journal of Diseases in Children*, 1922, Vol. 23, p. 392.

[11] Talbot FB. Eczema in childhood. *Medical Clinician North America*, 1918, Vol. 1, p. 985.

[12] Cogswell JJ, Alexander J. Breast feeding and eczema/asthma (letter). *The Lancet*, 1982, Vol. i, p. 910.

[13] Businco L, Marchetti F and Pellegrini G. Prevention of atopic diseases in 'at risk newborns' by prolonged breast-feeding. *Annals of Allergy*, 1983, Vol. 51, p. 296.

[14] Businco L, Cantani A, Meglio P and Bruno G. Prevention of atopy: results of a long-term (seven months to eight years) follow-up. *Annals of Allergy*, 1987, Vol. 59, pp. 183–186.

[15] Saarinen UM, Kajosaari M and Backman A. Prolonged breast-feeding as prophylaxis for atopic disease. *The Lancet*, 1979, Vol. 2, p. 63.

[16] Kajosaari M and Saarinen UM. Prophylaxis of atopic disease by six months total solid food elimination. *Acta Paediatrica Scandinavica*, 1983, Vol. 72, p. 411.

[17] Hamburger RN. Diagnosis of food allergy and intolerances in the study of prophylaxis and control groups in infants. *Annals of Allergy*, 1984, Vol. 53, p. 673.

[18] Gerrard JW. Allergy in breast-fed babies to ingredients in breast milk. *Annals of Allergy*, 1979, Vol. 42, p. 69.

[19] Warner JO. Food allergy in fully breast-fed infants. *Clinical Allergy*, 1980, Vol. 10, p. 133.

[20] Croner S and Kjellman N-IM. Predictors of atopic disease: cord blood IgE and month of birth. *Allergy*, 1986, Vol. 41, pp. 68–71.

[21] Bjorksten F and Suoniemi I. Dependence of immediate hypersensitivity and month of birth. *Clinical Allergy*, 1976, Vol. 6, pp. 165–171.

[22] Bjorksten F, Suoniemi I and Koski V. Neonatal birch-pollen contact and subsequent allergy to birch pollen. *Clinical Allergy*, 1980, Vol. 10, pp. 585–591.

[23] Morrison-Smith J and Spingett VH. Atopic disease and month of birth. *Clinical Allergy*, 1979, Vol. 9, pp. 153–157.

[24] Robert J and Carron R. Pollinose precose des natifs du Taureau. *Allergol Immunologie*, 1979, Vol. 19, pp. 153–155.

[25] Suoniemi I, Bjorksten F and Haahtela T. Dependence of immediate hypersensitivity in the adolescent period on factors encountered in infancy. *Allergy*, 1981, Vol. 36, pp. 263–268

[26] Quoix E, Bessot JC, Kopferschmitt-Kubler MC, Fraisse P and Pauli G. Positive skin tests to aero-allergens and month of birth. *Allergy*, 1988, Vol. 43, pp. 127–131.

[27] Schata M, Noebel A, Fabry H and Jorde H. Zusammehange zwischen Pollinosis und Geburtsmonat. *Allerologie*, 1983, Vol. 6, pp. 372–377.

[28] Robert J, Carron R and Bourgeois M. Pollinose et mois de naissance. *Bulletin of Active Therapy*, 1980, Vol. 80, pp. 451–453.

[29] Venables KM, Allitt U and Collier CG *et al*. Thunderstorm related asthma – the epidemic of 24/25th June 1994. *Clinical Experimental Allergy*, 1997, Vol. 27, pp. 725–736.

[30] Busse WW. The relationship between viral infections and onset of allergic diseases. *Clinical Experimental Allergy*, 1989, Vol. 19, pp. 1–9.

[31] Martinez FO. Role of viral infections in the inception of asthma and allergies during childhood: could they be 'protective'? Thorax, 1994, Vol. 49, pp. 1189–9131.

[32] Shaheen S. Discovering the causes of atopy: patterns of childhood infection and fetal growth may be implicated. *BMJ*, 1999, Vol. 314, pp. 987–988.

[33] Holt PG and Sly PD. Allergic respiratory disease: strategic targets for primary prevention in childhood. *Thorax*, 1997, Vol. 52, pp. 1–4.

[34] Shirakawa T, Enomoto T, Shimazu S-I and Hopkin JM. The inverse association between tuberculin responses and atopic disorders. *Science*, 1997, Vol. 275, pp. 77–79.

[35] Holgate ST. The cellular and mediator basis of asthma in relation to natural history. *The Lancet*, 1997, Vol. 350 (suppl. ii), pp. 5–9.

[36] Luczynska CM, Li Y and Chapman MD *et al*. Airborne concentrations and particle size distribution of allergen derived from domestic cats (*Felis domesticus*): measurements using cascade impactor, liquid impinger and a two site monoclonal antibody assay for Fel d 1. *American Review of Respiratory Disease*, 1990, Vol. 141, p. 361.

[37] Rudolph R, Meier-Duis H, Kunkel G, Staud RD and Stock U. Allergies to animal hair in diseases of the upper respiratory tract. *Dentsche Medizinische Wochenschrift*, 1975, Vol. 100, pp. 2557–2561.

[38] Rudolph R, Kunkel G, Blohm B *et al*. Frequency and clinical significance of sensitisation to animals. *Allergology*, 1981, Vol. 4, pp. 230–236.

[39] Bornehag CG, Sundell J, Hagerhed L, Janson S and the DBH study group. The 'healthy pet-keeping effect'. Pet-keeping in early childhood as a risk or a protection against allergic symptoms later in life. Conference Proceedings. The International Academy of Indoor Air Sciences, Monterey. *Indoor Air*, July 2002, Vol. III, pp. 398–403.

[40] Platts-Mills TAE, Custis N, Erwin EA, Sporik R and Woodfolk JA. Asthma and Indoor Air. The International Academy of Indoor Air Sciences, Monterey. *Indoor Air*, July 2002, Vol. III, pp. 10–16.

[41] Peat JK and Li J. Reversing the trend: reducing the prevalence of asthma. *Journal of Allergy and Clinical Immunology*, 1999, Vol. 103, pp. 1–10.

[42] Ronmark E, Jonsson E, Platts-mills TAE and Lundback B. Incidence and remission of asthma in schoolchildren: report from the obstructive lung disease in Northern Sweden studies. *Pediatrics*, 2001, Vol. 107, pp. E37.

[43] Perzanowski MS, Ronmark E, Platt-Mills TAE and Lundback B. The protective effect of having a cat or dog at home on development of asthma in pre-teen age children. *American Journal of Respiratory Critical Care Medicine*, 2002.

[44] Crater DD, Heise S, Perzanowski M, Herbert R, Morse CG and Platts-Mills TAE. Asthma hospitalization trends in Charleston, South Carolina from 1956–1997. *Pediatrics*, 2001, Vol. 108, pp. 1–6.

[45] Gergen PJ and Weiss KB. Changing patterns of asthma hospitalization among children: 1979 to 1987. *JAMA*, 1990, Vol. 264, pp. 1688–1692.

[46] Call RS, Smith TF, Morris E, Chapman M and Platts-Mills TA. Risk factors for asthma in inner city children. *Journal of Pediatrics*, 1992, Vol. 121, pp. 862–866.

[47] Rosenstrich DL, Eggleston P, Kattan M, Baker D, Slavin RG, Gergen P, Mitchell H, McNiff-Mortimer K, Lynn H, Ownby D and lalveaux F. The role of cockroach allergy and exposure to cockroach allergen in causing morbidity among inner-city children with asthma. *New England Journal of Medicine*, 1997, Vol. 336, pp. 1356–1363.

[48] Eggleston PA, Rosenstreich D, Lynn H, Gergen P, Baker D, Kattan M, Mortimer KM, Mitchell H, Ownby D, Slavin R and Malveaux F. Relationship of indoor allergen exposure to skin test sensitivity in inner-city children with asthma. *Journal of Allergy and Clinical Immunology*, 1988, Vol. 102, pp. 563–570.

[49] Sporik RB, Squillace SP, Ingram JM, Rakes G, Honsinger RW and Platts-Mills TAE. Mite, cat and cockroach exposure, allergen sensitisation and asthma in children. *Thorax*, 1999, Vol. 54, pp. 675–680.

[50] Blumenthal M, Bousquet J and Burney P. Evidence for an increase in atopic disease and possible causes. *Clinical Experimental Allergy*, 1993, Vol. 23, pp. 484–492.

[51] Mygind N, Dahl R and Peterson S. Thestrup-Pedersen K. Essential Allergy (2nd Edition). Blackwell, Oxford, 1996, ISBN 0–632–03645–1.

[52] Platts-Mills TAE. Dust mite allergens and asthma – a report of a second international workshop. *Journal of Allergy and Clinical Immunology*, 1992, Vol. 89, pp. 1046–1057.

[53] Bronswijk JEMH van and Sinha RN. Pyroglyphid mites (Acari) and house dust allergy: a review. *Journal of Allergy*, 1971, Vol. 47, pp. 31–52.

[54] National Asthma Campaign. Out in the open: A true picture of asthma in the United Kingdom today, *The Asthma Journal*, 2001, Vol. 6, 3 September.

[55] Blumenthal M, Bousquet J and Burney P. Evidence for an increase in atopic disease and possible causes. *Clinical Experimental Allergy*, 1993, Vol. 23, pp. 484–492.

[56] Jackson R, Sears MR, Beaglehole R and Rea H. International trends in asthma mortality: 1970 to 1985, *Chest*, 1988, Vol. 94, pp. 914–918.

[57] Lynden van Nes AMT. Effective mite allergen avoidance in households with asthmatic children, University press facilities, Eindhoven, 1999, ISBN 90–6814–097–3.

[58] Kitch BT, Chew G, Burge HA, Mulienberg Ml, Weiss ST, Platts-Mills TAE, O'Conner G and Gold DR. Socioeconomic predictors of high allergen levels in homes in the greater Boston area. *Environmental Health Perspectives*, 2000, Vol. 108, No. 4, pp. 301–307.

[59] Strachan DP, The role of environmental factors in asthma, *British Medical Bulletin*, 2000, Vol. 56, No. 4, pp. 865–882.

[60] Institute of Medicine, Clearing the Air, Committee on the Assessment of Asthma and Indoor Air, National Academic Press, Washington, 2002, ISBN 0–309–06496–1.

Chapter 2

The ecology and physiology of the house dust mite

One-fifth to a quarter of the European population are allergic to cats, mites or fungi.[1-4] The specific mite family 'Pyroglyphidae' have however been shown by Turos[5] to be of greater importance producing a large range of particularly allergenic proteins. Research by Korsgaard[6] into the faecal pellets excreted by the HDM has demonstrated that they have a direct causal and a dose–response relationship with asthma. Colloff et al.[7] have demonstrated that feeding on the abundance of dead skin scales shed by humans and animals, the only regulating factor in their common habitats is the availability of moisture. Wharton[8] has also shown that they have an ability to adapt and hibernate through non-optimum hygro-thermal conditions and only prolonged periods of low humidity or extremes in temperature can inhibit colonisation and limit proliferation rates.

The evidence, that they are the main causal factor in the current asthma pandemic in the UK, is becoming compelling. To evaluate whether an interventionist approach – which attempts to suppress HDM colonies and denature their allergens can prove effective, both as a treatment to ameliorate asthmatic symptoms, and as a long-term solution reducing the prevalence of the disease, requires an understanding of the physiology and ecology of the HDM and why it has become ubiquitous in the modern domestic environment.

Classifying the HDM

Fossil studies show that mites have existed on earth for over 400 million years, evolving 23 million years ago into scavengers living in birds nests or similar habitats. Walter[9] has estimated that there may be over 100 million different species of mites in existence, and – due to their adaptability – mite species can be found in almost every environment.

'Dust mite' is the generic term for a species classification forming a sub-set of a large family tree, phylum Arthropod, which accounts for up to 80% of all living creatures on the planet.[9] Due to the scale of this family,

HDM are associated with the arthropod branch of this tree, which descends further into arachnid and finally to the Acaridae family of mites and ticks. 'Arthropod' is a term for invertebrate animals having an external skeleton and jointed appendages, such as crustaceans, insects and spiders. It is believed there could be as many as half a million species of Acaridae, although Colloff[10] reported that only 30 000 have so far been officially discovered. The term 'House Dust Mite' applies to a taxonomically and ecologically defined subclass of all the species of mite commonly found in homes. Colloff[11] has reported that 13 species of mite have been found in house dust recorded from locations throughout the world, including the United States, Hawaii, Canada, Europe, Asia, the Middle East, Australia, South America and Africa. The Entomological Society of America (ESA) approved the common names for two of the most abundant species of mite found in house dust in Europe and North America respectively: the European house dust mite *Dermatophagoides* (Latin for skin eater) *pteronyssinus* (DP) – identified by Trouessart[12] in 1897 and the American house dust mite *Dermatophagoides farinae* (DF).[13] The only other commonly encountered species of this genus found in house dust are *Dermatophagoides microceras* and *Euroglyphus maynei*. A study by Blythe, Williams and Morrison[14] confirmed that these species of mite accounted for over 90% of the HDM fauna found in dwellings in Birmingham, UK. In general DF is more commonly found in North America and other regions with prolonged dry weather, whilst DP is abundant in areas with higher ambient humidity, such as the UK.[15]

Biology of the HDM

HDMs have seven separate phases in their life cycles. The development stages are: egg, prelarvae, larvae, protonymph, deutonymph, tritonymph and adult.[16] Mites exist in adulthood as female and male. Interaction between the two must occur for the purpose of reproduction.[17,18]

Arlain[19] has shown that DP females weigh approximately 5.8 μg whilst males are approximately half this weight at 3.5 μg. Between 72 and 74% of their total weight is water and due to their large surface area to volume ratio, the loss of water by evaporation is a problem which they minimise by having an exoskeleton and a variety of survival strategies. The ideal conditions for mites to proliferate is at a temperature of 25 °C and a relative humidity (RH) of 80%.[20] A high humidity is very important to the survival of these creatures as most of their water is gained from the atmosphere by osmosis. Mites live in an atmosphere where no liquid water exists and moisture balance is critical to their survival. A small gland by the first pair of legs leads to the mouth through a narrow groove or gutter. This gland secretes a salt solution. In damp conditions the salt solution, which is hygroscopic, absorbs water from the air down to a critical threshold level of 70–75% RH at 25 °C. In dry conditions, water evaporates from the solution so that the salts crystallise and block the opening of the gland. This slows down water loss. In prolonged dry atmospheres the mite eventually dries out and dies. Under ideal conditions the life span of a mite is approximately three months.[8]

Absolute humidity was initially thought to be the best guide to mite growth rates[16] and Platts-Mills and Chapman[15] initially recommended a mixing ratio of 9 g water vapour/kg dry air to be the critical point that determines mite survival. After further research this was reduced to 7 g water vapour/kg dry air.[21]

Critical equilibrium humidity

Crowther *et al.*[22] have maintained that it is RH that is more important. Mites have a high surface to volume ratio and are poikilothermic. As there is no difference in temperature between the mite and the environment, the relative difference between the air vapour pressure and the mite's internal saturation vapour pressure is proportional to relative, rather than absolute humidity. At a micro-climate level, mite survival does not appear to be dependent on absolute humidity. Relative humidity of 79% at 20 °C is the equivalent of 56% RH at 27 °C. Under the initial conditions mites survive and thrive; under the latter they die. Arlain[23] has claimed that the critical equilibrium humidity (CEH) for dust mites is 73% RH at 25 °C. Below the CEH, transpiration of water from the mite to the atmosphere is greater than the rate of absorption from the atmosphere, and dehydration occurs. At 40–50% RH individual adults survived for no longer than 8–11 days, however, a population consisting of all life stages survived for 10 weeks, but decreased in numbers. Increasing RH to above 85% may also be detrimental to mite populations as mould growth inhibits survival due to the production of toxins.[24,25] A fluctuating environment with periods of high RH to maintain water balance and low RH to minimise mould growth may thus be the preferred environment for mite colonisation and proliferation – a hygro-thermal regime typical of the domestic environment in the UK.

Mites are extremely resilient as they have more than one defence mechanism with which to combat periods of unsuitable conditions. They can migrate from areas where conditions are unsuitable; they can shutdown into a state similar to that of hibernation and also huddle together.[8] Arlain[26] has shown that these clusters consist of groups of 5–25 altering their surface to volume ratio to minimise water loss.

Adult mites, both male and female, mate more than once and during their active reproductive lives, female mites can produce 200–300 eggs. Mated females, under optimum conditions, lay eggs one at a time and can produce up to three eggs per day. The influence of environmental factors on the egg stage does not appear to be fully understood. It takes between 6 and 12 days to reach the hatching stage and DP eggs have also been shown to remain viable after seven months and 10 °C and 60% RH and may thus survive overwintering.[27] Prolonged optimum conditions are thus likely to produce large increases in the HDM population.

Conditions below 15 °C and above 35 °C slow down mite development.[8] Early stages of life are more resistant to extremes of climate, enabling them to survive dry periods and to re-emerge when favourable conditions return. De Boer, Kuller and Kahl[28] found HDM populations survived for 10 weeks at 16 °C when 1.5 hr of moist air (76% RH) was made available to them per day (remainder of the day at 35% RH) although egg production did not occur unless the period of moist air was increased to 3 hr. A population expansion did not occur until the period of moist air was increased to 6 hr per day. Reducing humidity also appeared to almost double the length of time it took to evolve from egg to breeding adult.

HDMs obtain water in four ways: ingestion of food, production of metabolic water from oxidation of carbohydrates and fats, passive absorption and active absorption from saturated air via salt glands. The main food of mites however is human skin. Humans shed approximately 0.5–1 g of skin per day,[29] which provides a considerable food reservoir. Korsgaard[30] has shown that several thousand mites are able to survive for months on just 0.25 g of food. The flakes of skin absorb moisture from the atmosphere and are colonised by yeast mould, genus *Aspergillus*.[31] The yeast causes the scales of skin to swell, moistening and softening them to aid digestion.[29] This adds further importance to the role of humidity in the lives of the HDM as moulds generally require an RH of 65% or greater to exist.[32] In addition mites can have a varied diet of pollen grains, bacteria and plant material and have also been shown to thrive on semen.[29] Humidity also plays a role in the quantity of food eaten. Below CEH mites consume less than 5% of their body weight per day compared with 10–51% above CEH. These higher feeding rates result in a proportionately greater output of faecal pellets.[33]

Ecology of the HDM

Ambient indoor conditions are less important to the mites than localised micro-climate, which is why high numbers of mites are commonly found in beds, particularly in seams and underneath buttons.[34] The reasons for this are that beds provide:

- A readily available and abundant food source – each human sheds between 0.5 and 1 g of dead skin every night.
- Moisture – through respiration and perspiration – somewhere in the region of 1 litre per night.
- Enclosed habitat – most beds are made in the morning, trapping in moisture and food. This maintains near constant conditions within the mattress.[35]

Field studies by Sesay and Dobson[36] and Bronswijk[37] have demonstrated that heavily used soft furnishings, especially mattresses and bedding, provide a supportive environment for the development of mite colonies. These findings support the conclusion that the older the mattress, the greater the HDM colony size. Through time the build-up of food and moisture create conditions which can approach the optimum. In 1995, Hay[38] vacuumed the surface area of a sprung mattress and recorded a population density of 3–46 living mites per square metre. This was three orders of magnitude below the level found when a core sample was taken from the upper 15 mm of the same mattress (8200–26 800 mites per square metre). When foam mattresses were compared with sprung mattresses, Abbott, Cameron and Taylor[39] found that they contained roughly similar numbers as a ratio of dust extracted (746 and 706 mites per gram, respectively). However the sprung mattresses contained over three times the weight of dust in foam mattresses and therefore the number of mites collected was $2489\,m^{-2}$ compared with $720\,m^{-2}$ for the other. This highlights the challenges associated with developing reliable measuring techniques. Carpet mite activity appears to be more variable and may be dependent on the type of carpet (wool/synthetic) and the depth of pile. Relative humidity has been shown to be 10% higher in carpets than in ambient air 1–2 metres above the floor.[23] This could be attributable to the decreased airflow from convection, combined with lower temperatures at floor level due to simple thermal stratification. Many fabrics are also hydroscopic and will absorb moisture from ambient air.

Mite numbers vary seasonally, rising and falling in accordance with the humidity cycle within the house.[16] Some studies have shown that the highest numbers of live mites experienced in Western Europe (including UK) occur in June, with the maximum numbers of dead mites found between August and September. Allergen counts reach peak levels between August and December, with the lowest level of both mite counts and allergen levels measured between February and May.[40,41]

Platts-Mills and Chapman[15] also found a highly significant increase in mite allergens from August to December, the highest concentrations being found in sofas. The author suggests that as mites avoid dry conditions, they will migrate away from the surface of furniture when ambient conditions cause drying to occur. Live mites will therefore only be present on the surface during periods of high ambient humidity. This migration into furniture is another reason why mite counting is inaccurate, as the number of mites captured is dependent on ambient conditions that are constantly varying. The rise in allergen levels occurred within one month of a recorded rise in humidity, however, the fall in allergen levels was delayed by as much as three months. The most likely reason given for this is that live mites persist in furniture and continue to produce allergen that comes to the surface in the form of faecal pellets, and that the reduction in mite numbers appears to reflect surface dehydration but major falls in the allergen production do not occur until all of the fabric has become dry, which may take several months. Cases of immediate hyper-sensitivity, common among patients with asthma, rhinitis and atopic dermatitis, are regarded as perennial because they do not exhibit strict seasonal exacerbations typical of pollinosis.[42,43]

Figure 2.1 is the thermographic external humidity data measured over a 24-month period in west central Scotland. It demonstrates that ambient absolute humidity remains below the critical $7\,g\,kg^{-1}$ dry air threshold, except for the period May to September.

Although RH is the main controlling factor in dust mite activity, the figure of $7\,g\,kg^{-1}$ of dry air at $21\,°C$ equates to an RH of 45%. If reasonable ventilation rates can be maintained during the winter months, when ambient humidity is at its lowest, there is a greater potential to progressively 'dry out' the HDM micro-climates ($3.5\,g\,kg^{-1}$ of dry air at $21\,°C$ equates to an RH of 22% – a figure well below the CEH).

Burr, Dean and Merrett[44] have shown that HDMs thrive in damp homes, but it is reasonable to suppose that this is because the moisture level in soft furnishings is influenced by the relative and absolute humidity of the room. Signs of home dampness, however, may not determine the presence of mite colonies as the soft furnishings in a dwelling may absorb excess air moisture, therefore sustaining micro-climates, without any visible signs of mould. If these micro-climates are maintained throughout the year, then no reduction in mite numbers – and their associated allergens – may be experienced and no change in symptoms is likely to occur.

Internal humidity will of course be influenced by moisture production and ventilation regimes. Well-heated but poorly ventilated dwellings with relatively tight construction techniques are likely to have high mixing ratios. When the heating system is in operation the RH may be low, however, the overnight cooling cycle experienced when the central heating system is

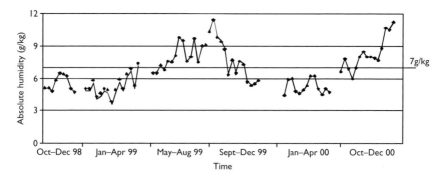

Figure 2.1 Absolute external humidity measured by on-site thermohygrograph.

switched off will result in a sharp rise in RH and a condensation cycle may result, wetting cold surfaces (typically windows or cold bridges) and permeating soft furnishings and carpets. The scope for drying HDM micro-climates may be greater during the winter months if a dwelling is reasonably well heated and ventilated.

HDM allergenic proteins

Kort and Kneist[45] have shown that HDM allergens are very robust and do not biodegrade. The combination of these factors means that exposure to this allergenic material can be frequent and prolonged. Dust mites are typically found in large numbers in beds and bedding where c. 8 hr every day are spent. At least 16 allergenic proteins have now been identified and characterised in HDM faecal pellets. Of these, Platts-Mills[46] maintains that the group I (*Der pI*) and group II (*Der pII*) allergens appear to be the most important, as more than 80% of mite allergic patients have developed IgE antibodies through regular exposure. A HDM produces 2–4 nanograms (ng) of group I allergen and about 1–2 ng of group II allergen per day.[10] The dry faecal pellets excreted are approximately 10–40 micro-millimetres in diameter[47] and are covered with a peritophic membrane (a membrane produced in the gut of the mite). Mites excrete approximately 20 pellets per day. Allergens are also contained in the skin of the mite,[9] which is shed twice in their lifetime. Such exuviae and bits of dead mite are also reactive. These allergenic parts are relatively large and will rapidly fall in undisturbed air,[48,49] however in time, the faecal pellets may be subject to further biodegradation and disturbance (depending on room activity) and the desiccated fragments – which will now be smaller and lighter – are likely to become more easily airborne and remain in suspension for longer periods.

The quantification of allergen exposure level is generally assessed on the ratio of *Der pI* to the amount of house dust found in a sample – µg *Der pI* per gram of house dust. This is the standard used by the World Health Organisation (WHO), who suggest that exposure to levels above a concentration of 2 Mg g, will result in sensitisation or induce allergic reactions. A further method of assessment has been the use of air sampling to produce absolute amounts of *Der pI* in the air in nanograms or ng m.$^{-3}$ There remains some debate over which method of sampling should be used to best indicate the risk of exposure to the patient. The former is quantifying the total reservoir burden available; the latter the quantity of aerosol allergen available in suspension capable of being respired. It may be useful to measure both over a period of time to assess any possible relationship, however, it is likely that the quantity of airborne allergen will be directly correlated with periods of activity and dust reservoir disturbance e.g. vacuuming, children playing, indoor aerobics etc.

Airborne allergen

Clark[50] demonstrated that household cleaning regimes were accompanied by a significant rise (10–15 fold) in the number of small airborne particles. Many of these particles were relatively heavy and resettled rapidly. Tovey *et al.*[51] suggest that very little allergen remains airborne for more than a few minutes although whole and fragmented mite bodies can become airborne during bed making.[52] Tovey *et al.*[51] also detected airborne allergen in disturbed air conditions but not in undisturbed air. This supports the findings of Swanson, Agarwal and Reed[53] who reported that concentrations of airborne allergen increased greatly during bed making. Sakaguchi *et al.*[54] examined airborne *Der pI* and *Der pII* under two distinct conditions; during everyday life in the living room and during bed making. Low levels were reported in the living room although large amounts were found on the floor, whilst in the bedroom, airborne allergen levels increased one-thousand fold. This suggests that while the size of allergen reservoir is important, it is of particular relevance during periods of human activity. A follow-up study by Sakaguchi *et al.*[55] found much higher airborne levels of mite allergen during sleep where subjects were using 'old' bedding. Airborne levels were ten times higher during sleep in the bedrooms than was previously found during everyday life in the living room. In addition Strachan and Carey[56] demonstrated that synthetic pillows had eight times higher *Der pI* levels than feather pillows. On this evidence Sakaguchi *et al.* advocate new bedding as an effective counter measure for reducing airborne allergens, as high levels of allergens are generated in bedding infested with mites. It may not take long – given optimum hygro-thermal conditions – for HDMs to colonise new bedding and generate a significant level of allergenic protein.

Two separate studies undertaken in 1996 by Peat *et al.*[57] and Custovic *et al.*[58] demonstrated a clear dose–response relationship between the amount of *Der pI* in the home and current asthma symptoms. Tovey *et al.*[51] maintained that an average dwelling house will contain about 1 kg of house dust which in turn will contain several milligrams of *Der pI*, picogram quantities of which are sufficient to create allergic lung reactions when inhaled. In terms of the proportion of mite bodies found to actual *Der pI* allergen, he suggested a likely ratio of 1 mite to 100 faecal pellets (based on live mites containing 1–1.5 ng of allergen and dead mites much less). Chapman and Platts-Mills[59] have demonstrated that a faecal sphere of around 20 microns in diameter, contains 0.1 ng of allergen. This means that the concentration of allergen in mite faeces is high, at approximately 10 mg *Der pI*/ml. Such a high concentration when combined with rapid elution rates – typical of allergic reactivity[60] – results in any particles reaching the lungs causing immediate, if localised, inflammatory responses.[61] Tovey, Chapman and Platts-Mills[62] suggest that the relationship between HDM allergy and bronchial asthma is best explained by faecal pellets having a cumulative inflammatory effect.

In provocation tests, the quantity of allergen entering the lung has been estimated to be the equivalent of one month's natural exposure.[60,63] Tovey et al.[80] suggest that for both pollen grains and dust particles, natural exposure represents a much slower rate of exposure than bronchial provocation, however, natural allergen is in a highly concentrated form and it seems likely that faecal pellets entering the lung would give rise to local inflammatory reactions. The study maintained that almost all airborne allergen were associated with particles greater than 10 microns in diameter and other studies have shown that particles above this size generally do not enter the lung.[51,52,63,64] It therefore follows that either the quantity of HDM allergenic proteins entering the lung is relatively low, or the pellets have been broken down into smaller fragments. Those that are deposited and absorbed by the lung wall are highly allergenic, leading to inflammation and progressive scarring in chronic asthmatics. Hence the constant wheeze associated with those that have had the condition for many years. Such lung damage reduces reversibility and inhibits the effectiveness of dilatory drug treatments.

Mite and allergen control

An effective allergy avoidance programme must consist of a set of measures to reduce exposure to relevant allergens or irritants, leading to a reduction of clinical symptoms. Any successful mite/allergen intervention will therefore involve three stages: eradicating live mites; denaturing the allergen reservoir and preventing – or at least inhibiting – re-colonisation. Zock and Brunekreef[65] showed that dust mites can rapidly re-colonise if environmental conditions are suitable.

Three controlled investigations of whole-house MHRV (mechanical heat recovery ventilation) systems – as a preventative measure in HDM allergy – have been conducted by Harving et al.[66–68] in Denmark. The studies examined the effect of patients allergic to HDMs moving into new dwellings which incorporated mechanical ventilation systems. Ventilation rates, indoor humidity, dust mites counts and clinical data were collected at regular intervals.

The results demonstrated an increase in ventilation rates from 0.3 to 1.3 ach^{-1} and produced a significant reduction in indoor humidity, as well as HDM allergen exposure. Subjective and objective clinical measurements demonstrated significant improvements in patients and these improvements were highly correlated with the changes in mite counts. Such a strategy can initially produce benefits from mite removal, alleviating symptoms in allergic individuals (secondary prevention) and secondly, may reduce the incidence of allergic disease in future populations (primary prevention).

The task is to test whether the benefits of such a proven approach can be replicated throughout the UK, where climatic variations in ambient relative and absolute humidity combine with poor house conditions to produce significant and influential variables and confounders.

References

[1] Kuehr J, Frischer T, Karmaus W, Meinert R, Barth R, Schraubs S, Daschner A, Urbanek R and Forster J. Natural variation in mite antigen density in house dust and relationship to residential factors. *Clinical Experimental Allergy*, 1994, Vol. 24, pp. 229–237.

[2] Popp W, Rauscher H, Sertl K, Wanke T and Zwick H. Risk factors for sensitization to furred pets. *Allergy*, 1990, Vol. 45, pp. 75–79 and 1988, Vol. 43, pp. 127–131.

[3] Desjardins A, Benoit C, Ghezzo H, L'Archeveque J, Leblanc C, Paquette L, Cartier A and Malo JL. Exposure to domestic animals and risk of immunologic sensitisation in subjects with asthma. *Journal of Allergy and Clinical Immunology*, 1993, Vol. 91, pp. 979–986.

[4] Weeke ER. Epidemiology of allergic disease in children. *Rhinology Supplement*, 1992, Vol. 13, pp. 5–12.

[5] Turos M. Mites in house dust in the Stockholm area. *Allergy*, 1979, Vol. 34, pp. 11–18.

[6] Korsgaard J. Mite asthma and residency. *American Review Respiratory Disease*, 1983, Vol. 128, pp. 231–235.

[7] Colloff M, Ayres J, Carwell F, Howarth P, Merrett TG, Mitchell EB, Walshaw M, Warner JO, Warner JO and Woodcock AA. The control of allergens of house dust mites and domestic pets: a position paper. *Clinical Experimental Allergy*, 1992, Vol. 22 (suppl. 2), pp. 1–28.

[8] Wharton GW. House dust mites. *Journal of Medical Entomology*, 1976, Vol. 12, pp. 577–621

[9] Walter DE. *Mites: Ecology, Evolution and Behaviour*, 1999, ISBN 08199–375.

[10] Colloff M. Differences between the allergen repertoires of house dust mites and stored product mites. *Clinical Immunology*, 1993, Vol. 16, p. 2.

[11] Collof MJ. Taxonomy and identification of dust mites. *Allergy*, 1998, Vol. 53 (suppl. 48), pp. 7–17.

[12] Trouessart EL. Acari Myriopoda et Scorpiones Hucusque in Italia Reperta. Patavii I. Cryptostigmata. 1897, Fasc 92.

[13] Wressell HB. Proposed additions to the list of common names on insects 1970. *Bulletin of the Entomological Society of America*, 1974, Vol. 20, p. 212.

[14] Blythe ME, Williams JD and Morrison SJ. Distribution of pyroglyphid mites in Birmingham with particular reference to *Euroglyphus maynei*. *Clinical Allergy*, 1974, Vol. 4, pp. 25–33.

[15] Platts-Mills TAE and Chapman MD. Dust mites: immunology, allergenic disease and environmental control. *The Journal of Allergy and Clinical Immunology*, 1987, Vol. 80, No. 6, pp. 755–775.

[16] Spieksma F Th M, Spieksma-Boezeman MIA. The mite fauna of house dust with particular reference to the house dust mite *Dermatophagoides pteronyssinus* (Trouessart 1897). *Acarologia*, 1967, Vol. 9(1), pp. 226–241.

[17] Larson DG. The critical equilibrium activity of adult females of the house dust mite *Dermatophagoides farinae*. PhD Dissertation. 1969, Columbus, Ohio State University

[18] Furumizo RT. The biology and ecology of the house dust mite *Dermatophagoides* farinae, Hughes 1961 (Acarina: Pyroglyphidae). PhD Dissertation, 1973, University of California.

[19] Arlian LG. Biology and ecology of house dust mites *Dermatophagoides* spp. and *Euroglyphus* spp., *Immunological Allergy Clinician North America*, 1989, Vol. 9, pp. 339–356.

[20] Hallas TE. The biology of mites. *Allergy*, 1990, Vol. 11, pp. 6–9.

[21] Platts-Mills TAE and Weck AL de. Dust mite allergens and asthma – a worldwide problem. *Journal of Allergy and Clinical Immunology*, 1989, Vol. 83, pp. 416–427.

[22] Crowther D, Horwood J Baker N, Thomson D. Pretlove S, Ridley I and Oreszcyn T. *House Dust Mites and the Built Environment: A Literature Review*, working document supporting an ESPRC Project. 2000, The Bartlett, UCL, London, September.

[23] Arlian LG. Water balance and humidity requirements of house dust mites. *Experimental and Applied Acarology*, 1992, Vol. 16, pp. 15–35.

[24] Arlain LG, Neal JS and Bacon SW, Survival, fecundity and development of DF at fluctuating relative humidity, *Journal of Medical Entomology*, 1998, Vol. 35, No. 6, pp. 962–966.

[25] Asselt L Van. Interactions between domestic mites and fungi. *Indoor and Built Environment*, 1999, Vol. 8, pp. 216–220.

[26] Arlain LG, Water exchange and effect of water vapour activity on metabolic rate in the dust mite *Dermataphagoides*. *Journal of Insect Physiology*, 1975, Vol. 21, pp. 1439–1442.

[27] Collof MJ. Effects of temperature and relative humidity on development times and mortality of eggs from laboratory and wild populations of the European house dust mite *Dermataphagoides pteronysinus*, *Experimental and Applied Acarology*, 1987, Vol. 3, pp. 279–289.

[28] De Boer R, Kuller K and Kahl O. Water balance of *Dp* maintained at briefly daily spells of elevated air humidity. *Journal of Medical Entomology*, 1998, Vol. 35(6), pp. 905–910.

[29] Whitrow D and Pycock R (eds). House dust mites: how they affect asthma, eczema and other allergies. Elliot Right Way Books, 1993, Surrey, England.

[30] Korsgaard J. Epidemiology of house dust mites. *Allergy*, 1998, Vol. 53 (suppl. 48), pp. 36–40.

[31] Douglas AE and Hart BJ. The significance of the fungus *Aspergillus penicilloides* to the house dust mite *Dermatophagoides pteronyssinus*, 1989, *Symbiosis*, Vol. 7, pp. 105–117.

[32] Hart BJ, Whitehead L. Ecology of house dust mites in Oxfordshire. *Clinical Experimental Allergy*, 1990, Vol. 20, pp. 203–209.

[33] Arlian LG. Humidity as a factor regulating feeding and water balance of the house dust mites *Dermatophagoide farinae* and *D. pteronyssinus* (Acari: Pyroglyphidae). *Journal of Medical Entomology*, 1977, Vol. 14, pp. 484–488.

[34] Mollet JA. Dispersal of American house dust mites (Acari: Pyroglyphidae) in a residence. *Journal of Medical Entomology*, 1996, Vol. 33, No. 5, pp. 844–847.

[35] Hay DB. *Ecology of the House Dust Mite*. Linacre College, Oxford University Press, 1991.

[36] Sesay HR and Dobson RM. Studies on the mite fauna of house dust in Scotland with special reference to that of bedding. *Acarologia*, 1972, Vol. 14, pp. 384.

[37] Bronswijk JEMH van. *Dermatophagoides pteronyssinus* (Trouessart 1897) in mattress and floor dust in a temperate climate (Acari: Pyroglyphidae). *Journal of Medical Entomology*, 1973, Vol. 10, pp. 63–70.

[38] Hay DB. An *in situ* coring technique for estimating the population size of HDMs in their natural habitat. *Arcologia*, 1995, Vol. 36, No. 4, pp. 341–345.

[39] Abbott J, Cameron J and Taylor B. House dust mite counts in different types of mattresses, sheepskins and carpets and a comparison of brushing and vacuuming collection methods. *Clinical Allergy*, 1981, Vol. 11, pp. 589–595.

[40] Platts-Mills TAE, Hayden ML, Chapman MD and Wilkins SR. Seasonal variation in dust-mite and grass pollen allergens in dust from the houses of patients with asthma. *Journal of Allergy and Clinical Immunology*, 1987, Vol. 79, pp. 781–791.

[41] Vervloet D, Penaud A and Razzouk H *et al*. Altitude and house dust mites. *Journal of Allergy and Clinical Immunology*, 1982, Vol. 69, pp. 290–296.

[42] Kern A. Dust sensitisation in bronchial asthma. *Medical Clinician of North American*, 1921, Vol. 5, p. 751.

[43] Tuft LA. Importance of inhalant allergens in atopic dermatitis. *Journal Invest Dermatology*, 1949, Vol. 12, p. 211.

[44] Burr ML, Dean BV and Merrett TG. Effects of anti-mite measures on children with mite-sensitive asthma: a controlled trial. *Thorax*, 1980, Vol. 35, pp. 506–512.

[45] Kort HSM and Kneist FM. Four-year stability of *Der PI* in house dust under simulated domestic conditions *in vitro*. *Allergy*, 1994, Vol. 49, pp. 131–133.

[46] Platts-Mills TAE. and Mitchell EB. House dust avoidance. *Lancet*, 1982, Vol. 2, pp. 1334–1336.

[47] Bronswijk JEMH van. House dust biology for allergists, acarologists and mycologists. Zoelmond, 1981.

[48] Platts-Mills TAE, Ward GW, Sporik R, Gelber LE, Chapman MD and Heymann PW. Epidemiology of the relationship between exposure to indoor allergens and asthma. *International Archives of Allergy and Immunology*, 1991, Vol. 94, pp. 339–345.

[49] Platts-Mills TAE, Thomas W, Aalberse RC, Vervloet D and Chapman MD. Dust mite allergens and asthma: report of a second international workshop. *Journal of Allergy and Clinical Immunology*, 1992, Vol. 89, pp. 1046–1060.

[50] Clark RP. Skin scales among airborne particles. *Journal of Hygiene* (Camb), 1974, Vol. 72 (10), pp. 47–51.

[51] Tovey ER, Chapman MD, Wells CW and Platts-Mills TAE. The distribution of dust mite allergen in the houses of patients with asthma. *American Review of Respiratory Disease*, 1981, Vol. 124, pp. 630–635.

[52] Harper GJ and Morton JD. The respiratory retention of bacterial aerosols: experiments with radioactive spores. *Journal of Hygiene* (Camb), 1953, Vol. 51, pp. 372–385.

[53] Swanson MC, Agarwal MK and Reed CE. An immunochemical approach to indoor aeroallergen quantitation with a new volumetric air sampler: studies with mite, roach, cat, mouse and guinea pig antigens. *Journal of Allergy and Clinical immunology*, 1985, Vol. 76, pp. 724–729.

[54] Sakaguchi M, Inouye S, Yasueda H, Irie T, Yoshizawa S and Shida T. Measurement of allergens associated with dust mite allergy. II. Concentrations of airborne mite allergens (*Der I* and *Der II*) in the house. *International Arch Allergy and Immunology*, 1989, Vol. 90, pp. 190–193.

[55] Sakaguchi M, Inouye S, Yasueda H and Shida T. Concentration of airborne mite allergens (*Der I* and *Der II*) during sleep. *Allergy*, 1992, Vol. 47, pp. 55–57.

[56] Strachan D and Carey, I. Reduced risk of wheezing in children using feather pillows is confirmed. *BMJ*, 1997, Vol. 314, p. 518.

[57] Peat JK, Tovey E, Toelle BG, Haby MM, Gray EJ, Mahmic A *et al.* House dust mite allergens. A major risk factor for childhood asthma in Australia. *American Journal of Respiratory Critical Care Medicine*, 1996, Vol. 153, pp. 141–146.

[58] Custovic A, Taggert S, Francis H, Chapman M and Woodcock A. Exposure to house dust mite allergens and the clinical activity of asthma. *Journal of Allergy and Clinical Immunology*, 1996, Vol. 98, pp. 64–72.

[59] Chapman MD and Platts-Mills TAE. *Journal of Immunology*, 1980, Vol. 125, pp. 587–592.

[60] Marsh DG. Allergens and the genetics of allergy. In: M. Sela (ed.) *The Antigens*, New York Academic Press, 1975, pp. 271–359.

[61] Austen FK and Orange RP. *American Review of Respiratory Disease*, 1975, Vol. 112, pp. 423–436.

[62] Tovey ER, Chapman MD and Platts-Mills TAE. Mite faeces are a major source of house dust allergens. *Nature*, 1981, Vol. 289, p. 592.

[63] Bates DV, Fish BR, Hatch TF, Mercer TT, Morrow PE. Deposition and retention models for internal dosimetry of human respiratory tract. *Health Physics*, 1966, Vol. 12, pp. 173–178.

[64] Wilson AF, Novey MS, Berke RA and Suprenant EL. Deposition of inhaled pollen and pollen extract in human airways. *New England Journal of Medicine*, 1973, Vol. 288, pp. 1056–1059.

[65] Zock JP and Brunekreef B. House dust mite allergen levels in dust from schools with smooth and carpeted classroom floors. *Clinical Experimental Allergy*, 1995, Vol. 25, pp. 540–553.

[66] Harving H, Korsgaard J. and Dahl R. Mechanical ventilators in dwellings as a protective measure in mite asthma. *Allergy and Asthma Proceedings*, 1988, Vol. 9: 283 (abstract 139).

[67] Harving H, Hansen LG, Korsgaard J, Nielsen PA, Olsen OF, Romer J, Svendsen UG and Osterballe O. House dust mite allergy and anti-mite measures in the indoor environment. *Allergy*, 1991, Vol. 46 (suppl. 11), pp. 33–38.

[68] Harving H, Korsgaard J and Dahl, R. House dust mite exposure reduction in specially designed, mechanically ventilated 'healthy' homes. *Allergy*, 1994, Vol. 49, pp. 713–718.

Chapter 3

The indoor environment and the house dust mite

Within the scientific research community the primary causal factors for the increasing prevalence of asthma remain contentious. Pollution, changes in diet, lifestyle, reduction in breast feeding, paracetamol, prophylactic and/or regular use of antibiotics, and a more sterile domestic environment have all been hypothesised, as has over-reporting due to a greater public awareness of health-related issues. Studies by Burr et al.,[1] Rona Chinn and Burney[2] and Omran and Russell[3] suggest that the prevalence of asthma has risen and cannot be wholly explained by greater awareness. At least part of the increase appears to be real. Austin et al.[4] in a study of asthma in the Scottish highlands, and Strachan et al.[5] in a national survey both concluded that macro-environmental factors, such as outdoor pollution and climate – although known triggers of respiratory symptoms[6,7] – were unlikely to be responsible for the rise in the prevalence of asthma. Both studies reported similar rates between natives and migrants in rural and urban areas, although the severity and frequency of attacks were higher in urban areas. This trend was also paralleled with decreasing socio-economic status.

The dust mite hypothesis

The HDM and the specific mite family 'Pyroglyphidae' are now thought to be of particular importance.[8] The genus *Dermatophagoides* and in particular the two species, *Dermatophagoides pteronyssinus* and *Dermatophagoides farinae* are the most abundant mites in Western Europe and North America respectively.[9,10] Maunsell, Wraith and Cunnington[11] and Miyamoto et al.[12] have claimed that these mites are known to be a key factor in the development of allergic disease (Figure 3.1).

Morrison-Smith et al.[13] estimated that around 80% of current asthmatics in the UK react to extracts of mite allergens, making the HDM and its faecal products one of the most ubiquitous environmental sensitisers and triggers. Esmen[14] has also claimed that contemporary leisure pursuits provided by

Figure 3.1 Portrait of *Dermatophagoides pteronyssinus.*

television and computers have increased time spent indoors, resulting in prolonged exposure to higher levels of airborne indoor allergens.

Further evidence strongly implicating the HDM as a prime causal factor came from an investigation into the rise in the prevalence of asthma in Papua New Guinea by Dowse *et al.*[15] The work demonstrated that the HDM was a significant factor in the pathogenesis of the condition. The tribe historically had very low rates of asthma, however after missionaries delivered a consignment of second-hand cotton blankets from Australia, a dramatic rise in asthma prevalence occurred in a relatively short space of time. The blankets were rarely washed and the research confirmed high infestation rates. It was estimated that 91% of the asthmatics had significant skin test reactions to extracts of *Dermatophagoides* mites and had grossly elevated serum-IgE antibody levels to the same species. Unusually the asthmatic symptoms were more prevalent in adults and the incidence of symptoms was 46 times that of a neighbouring tribe that used traditional bedding materials. As other factors which are normally implicated (increased hygiene, breast feeding, drug use, immunisation, pollution and diet) remained relatively unchanged over the study period among both tribes, such findings lend considerable weight to the HDM hypothesis as constituting an independent and significant driver in the aetiology of the disease.

The ideal condition for mites to proliferate is a RH of 80% at 25 °C.[16] A high humidity is very important to the survival of these creatures as most of their water is gained from the atmosphere by osmosis.[17] The mites live in an environment where no liquid water exists and moisture balance is critical to their survival. Under ideal conditions the life span of a mite is two to three months.[18] Mite numbers vary seasonally, rising and falling in accordance with the humidity cycle within the house.[19,20] The highest numbers are experienced in middle to late summer, when ambient RH is usually at its highest.[21]

As the main food of mites is human skin, heavily used soft furnishings, mattresses and bedding provide a perfect environment for the establishment and proliferation of mite colonies.[22,23] The flakes of skin absorb moisture from the atmosphere and are colonised by a mould.[24] This causes the skin scales to swell,[25] moistening and softening them to aid digestion. As Gravesen[26], Hart and Whitehead[27] have demonstrated that moulds generally require an RH of 65% or greater to exist, hence a warm, humid house provides an ideal habitat.

HDM allergens: sensitisation thresholds

An association between house dust and allergic reactions was demonstrated in 1928 by Dekker.[28] In 1967, Voorhorst *et al.*[29] identified the species *Dermatophagoides* as a source of allergen capable of inducing allergic reactions. By 1981 Tovey *et al.*[30] had discovered that individual faecal pellets contain very high concentrations of an allergenic protein and particles, and on reaching the lungs could trigger allergic reactions in susceptible individuals. Furthermore, in 1983 Korsgaard[31] demonstrated that there was a clear dose–response relationship between exposure levels to allergenic particles and disease prevalence. In a review of literature on the subject of mites and their allergens, Sporik, Chapman and Platts-Mills[32] indicated that the evidence strongly suggested that a causal relationship exists between exposure to mite allergens and asthma. The world of entomology and immunology have been well aware of the HDM as a causal factor in lung disease for over 20 years.

This growing importance in the role of the HDM and its association with allergic diseases prompted the WHO to design guidelines for the control of dust mites. After investigation by Platts-Mills and de Weck[20] a figure for absolute humidity of $7\,g\,kg^{-1}$ of dry air was stipulated as the limiting factor for the growth of colonies. Below this level numbers of mites begin to fall due to direct desiccation of the mites themselves, plus the dehydration of the skin scales on which they feed.

Given optimum conditions, dust mites live for two months. During their active reproductive life, females were observed by Furumizo[33] to produce – under laboratory conditions – 200–300 eggs. It is thus theoretically possible for one pair of dust mites – given optimum environmental conditions – to generate over 200 million offspring in a 12-month period, based on the assumption of an equal gender balance. As Tovey *et al.*[30] demonstrated that each dust mite can produce up to 60 times its own body weight in faecal pellets during its life span, resulting in a cumulative burden of allergens building in the micro-climates. Furthermore, allergens produced by the HDMs do not appear to decay naturally and have been shown by Kort and Kneist[34] to be stable for at least four years. Even accepting that optimum conditions will not necessarily prevail for the majority of any given time frame, the geometric growth of dust mite colonies have the potential to produce large bio-cumulative, reservoirs of allergen.

The recent development of an assay to identify *Der p1* – one of the fifteen allergenic proteins which have been identified in dust mite faeces – can indicate and quantify the historical level of dust mite activity. The assay provides outputs in micrograms per gram of house dust or as absolute weights per square metre. The results are a reliable longitudinal marker to assess the historical level of dust mite activity in any particular dwelling. In 1987 Platts-Mills and de Weck[20] suggested safety limits for exposure to dust mites and their allergens of 100 mites per gram of dust, or 2 mg *Der p1* per gram of house dust, as a level above which prolonged exposure increases the risk of sensitisation and the production of mite-specific antibodies in atopic individuals. If a level of 500 mites or 10 mg *Der p1* per gram of dust is exceeded then there is an increased risk of a severe allergic reaction in genetically 'at risk' individuals. Keuhr *et al.*[35] have claimed that for non-atopic individuals, the much higher level threshold of 60 µg *Der p1/g* of house dust, is a more appropriate sensitising threshold. What is important here is the realisation that almost everybody is at risk of developing asthma if exposed to high levels of HDM proteins. It is more than likely that certain construction techniques and living patterns will result in greater indoor concentrations developing through time. A large scoping study is urgently required to assess domestic HDM burdens across various house types.

Although useful as comparative markers, only the allergen in suspension will enter the lungs. More research is also required into the relationship between human activity, allergen reservoir disturbance and particulate mass and size and how this reacts with the lungs.

Housing and dampness

Several epidemiological studies[36–39] have identified the health implications of living in damp homes. Dampness has been strongly associated with respiratory disease and in particular asthma.[40–44] The 1996 Scottish House Condition Survey[45] established that 25% of all dwellings suffer from problems of dampness or condensation. Persistent condensation dampness, which will usually form on the external corner of thermally isolated bedrooms, will inevitably lead to the growth of mould fungal spores which thrive on this distilled water. (NB it is rare to find mould on rising dampness as the ground salts can inhibit mould growth.) *Aspergillus* is one of the most common moulds found in dwellings suffering from condensation. *Aspergillus* appears to play an important role in rendering the skin follicles edible[24] to the HDM. As this mould is only found in temperate northern or southern latitudes, it provides an explanation as to why the increase in asthma prevalence in such latitudes, is not being paralleled in tropical or equatorial regions.

Assessing allergen eradication techniques

A recent meta-analysis undertaken by Gotzsche, Hammerquist and Burr[46] published in the British Medical Journal concluded that current physical mite eradication techniques are ineffective and would therefore offer little benefit to asthma sufferers. The review identified 429 papers that held some relevance in the field of asthma and HDMs. When assessed against their chosen criteria, 24 completed studies were examined in detail. Of these trials, 18 failed to reduce dust mite levels – either through inappropriate methods or improper application. Of the completed studies examined in the paper, only six succeeded in reducing dust mite/allergen levels, and in these studies, a positive response was demonstrated.

Walshaw and Evans[47] selected 50 adult asthmatics, with strongly positive skin prick tests to HDM extracts, and attacked bedroom allergen reservoirs. Mattresses were encased, bedding renewed and a weekly cleaning programme advocated to participants. Where possible, carpets and all unnecessary soft furnishings were removed. The study concluded that successful house dust eradication procedures are possible and that patients allergic to the HDM appeared to have both subjective and objective improvements in their asthma. There is also a suggestion that a longer trial incorporating mattress protectors would be of value.

A similar project by Dorward et al.,[48] which included the initial killing of mites with liquid nitrogen, also demonstrated improvements in 21 adult asthmatic patients and concluded that the initial killing of mites should be routinely undertaken as an essential part of any future study. The study did not measure Der pI allergen levels before and after the intervention or assess whether liquid nitrogen had any effect on the existing reservoirs.

Enhert et al.[49] split 24 asthmatic children into three groups where one group used an acaricide on bedroom carpets and mattresses, a second used a placebo acaricide and the third utilised an encasing regime. Only the third strategy, which involved encasing of mattresses, duvets and pillows, produced any significant reduction in dust mite allergen levels and indicated that this had some positive results in terms of reduced bronchial hyper-reactivity. Similar methods and results were found by Carswell et al.[50] who indicated that such an approach may only be of benefit to highly mite-sensitive asthmatics. A high proportion of asthmatic children, however, test positive to HDM proteins.

Warner, Marchant and Warner[51] supplied ionisers to 20 asthmatic children to be used in living rooms during the day and bedrooms at night. Although the amount of airborne allergen was reduced, there was no improvement in the participants' symptoms. To be effective the allergen has to be airborne. The ionisers, however, were only being run when disturbance to reservoirs was at a minimum. The study also considered the importance of prolonged low exposure to airborne allergens versus short-duration high exposure, such as inhaling high concentrations from the pillow. The final study that reduced allergen levels argued that the education of any participants in dust mite avoidance techniques was vital for a successful prevention regime. Huss et al.[52] split 52 adult asthmatics into two groups where one group received conventional instructions (counselling and written instructions) and the other received conventional instructions plus a 22-minute interactive computer-based demonstration. Both groups lowered allergen levels on the mattresses over the 12-week monitoring period, with only the latter group achieving lower dust mite levels on bedroom carpets and floors. Bronchodilator use was reduced in the computer-assisted group, however, no other symptoms improved.

The conclusion of the meta-analysis appeared to be at odds with the evidence uncovered. Where eradication techniques were effective, asthmatic symptoms were reported to improve to varying degrees. The meta-analysis is thus more a critique of the inadequacies of single-strategy applications, and points to the need for a more holistic and comprehensive allergen denaturing and suppression regime to be tested.

The benefits of an allergen-free environment

Spieksma et al.[19] and Charpin et al.[53] demonstrated the benefits that can be gained from a dust mite and allergen-free environment. Their research projects conducted at high altitude – where the air is very cold and thus has a low moisture content – resulted in very little dust mite activity. When asthmatic children were moved to a commune in the mountains as a treatment for asthma, within two to three months, improvements in almost all asthma symptoms were reported. Similar results were demonstrated in a hospital trial by Platts-Mills and Mitchell[54] where – apparently due to factors such as mattress encapsulation, hard floor surfaces and cleaning regimes – there was little dust mite activity. These studies successfully demonstrated that asthmatic patients living in a relatively allergen-free environment can symptomatically improve. It is thus apparent from this literature review that any study designed to investigate the role of HDM on asthmatic patients must involve not only the inhibition of mite colonies, but should also include an effective strategy to attack the existing reservoir of dust mite allergens contained in a variety of micro-climates.

References

[1] Burr ML, Butland BK, King S and Vaughan-Williams E. Changes in asthma prevalence: two surveys 15 years apart. *Archives of Disease in Childhood*, 1989, Vol. 64, pp. 1452–1456.

[2] Rona RJ, Chinn S and Burney PGJ. Trends in the prevalence of asthma in Scottish and English primary school children 1982–92. *Thorax*, 1993, Vol. 50 pp. 992–993.

[3] Omran M and Russell G. Continuing increase in respiratory symptoms and atopy in Aberdeen schoolchildren. *BMJ*, 1996, Vol. 312, p. 34.

[4] Austin JB, Russell G, Adam MG, Mackintosh D, Kelsey S and Peck DF. Prevalence of asthma and wheeze in the Highlands of Scotland. *Archives of Disease in Childhood*, 1994, Vol. 71, pp. 211–216.

[5] Strachan DP, Anderson HR, Limb ES, O'Neill A and Wells N. A national survey of asthma prevalence, severity and treatment in Great Britain. *Archives of Disease in Childhood*, 1994, Vol. 70, pp. 174–178.

[6] Andrae S, Axleson O, Bjorksten B, Fredriksson M and Kjellman N-IM. Symptoms of bronchial hyper-reactivity and asthma in relation to environmental factors. *Archives Disease in Childhood*, 1988, Vol. 63, pp. 473–478.

[7] Abramson M and Voight T. Ambient air pollution and respiratory disease. *Medical Journal of Australia*, 1991, Vol. 154, pp. 543–551.

[8] Turos M. Mites in house dust in the Stockholm area. *Allergy*, 1979, Vol. 34, pp. 11–18.

[9] Blythe ME. Some aspects of the ecological study of the house dust mite. *British Journal of Chest Disease*, 1976, Vol. 70, pp. 3–21.

[10] Burgess I. Allergic reaction to arthropods. *Indoor Environment*, 1993, Vol. 2, pp. 64–70.

[11] Maunsell K, Wraith DG and Cunnington AM. Mites and house dust allergy in bronchial asthma. *The Lancet*, 1968, Vol. 1, pp. 1267–1270.

[12] Miyamoto T, Oshima S, Ishizaki T and Sato S. Allergic identity between the common floor mite (*Dermatophagoides farinae*) and house dust as a causative antigen in bronchial asthma. *Journal of Allergy*, 1968, Vol. 42, pp. 14–28.

[13] Morrison-Smith J, Disney ME, Williams JD and Goels ZA. Clinical significance of skin reactions to mite extracts in children with asthma, *BMJ*, 1969, Vol. 2, pp. 723–726.

[14] Esmen NA. The status of indoor pollution. Environmental Health Perspectives, 1985, Vol, 62, pp. 259–265.

[15] Dowse GK, Turner KJ, Stewart GA, Alpers MP and Woolcock AJ. The association between Dermatophagoides mites and the increasing prevalence of asthma in village communities within the Papau New Gineau Highlands. *Journal of Allergy and Clinical Immunology*, 1985, Vol. 75, pp. 75–83.

[16] Arlian LG. Biology and ecology of house dust mites *Dermatophagoides* spp. and *Euroglyphus* spp. *Clinical Immunology and Allergy*, North America, 1989, Vol. 9, pp. 339–356.

[17] Hallas TE. The biology of mites. *Allergy*, 1990, Vol. 11, pp. 6–9.

[18] Wharton GW. House dust mites. *Journal of Medical Entomology*, 1976, Vol. 12, pp. 577–621.

[19] Spieksma FTH, Voorhorst R, Varekamp H, Leupen MJ and Lyklema AW. The house dust mite (*Dermatophagoides pteronyssinus*) and the allergens it produces: Identity with the house dust allergen. *Journal of Allergy*, 1967, Vol. 39, pp. 325.

[20] Platts-Mills TAE and de Weck AL. Dust mite allergens and asthma – a worldwide problem. *Journal of Allergy and Clinical Immunology*, 1989, Vol. 83, pp. 416–427.

[21] Bronswijk JEMH van and Sinha RN. Pyroglyphid mites (Acari) and house dust allergy: A review. *Journal of Allergy*, 1971, Vol. 47, pp. 31–52.

[22] Sesay HR and Dobson RM. Studies on the mite fauna of house dust in Scotland with special reference to that of bedding. *Acarologia*, 1972, Vol. 14, pp. 384.

[23] Bronswijk JEMH van. *Dermatophagoides pteronyssinus* (Trouessart 1897) in mattress and floor dust in a temperate climate (Acari: Pyroglyphidae). *Journal of Medical Entomology*, 1973, Vol. 10, pp. 63–70.

[24] Douglas AE and Hart BJ. The significance of the fungus *Aspergillus penicilloides* to the house dust mite *Dermatophagoides pteronyssinus*. *Symbiosis*, 1989, Vol. 7, pp. 105–117.

[25] Whitrow D. *House Dust Mites: How They Affect Asthma, Eczema and Other Allergies*. Elliot Right Way Books, 1993, Surrey, England.

[26] Gravesen S. Fungi as a cause of allergic disease. *Allergy*, 1979, Vol. 34, pp. 135–154.

[27] Hart BJ and Whitehead L. Ecology of house dust mites in Oxfordshire. *Clinical Experimental Allergy*, 1990, Vol. 20, pp. 203–209.

[28] Dekker H. *Asthma und milben. Munch Med Wochenschr*, 1928, Vol. 75, pp. 515–516.

[29] Voorhorst R, Spieksma FTH, Varekamp H, Leupen MJ and Lyklema AW. The house dust mite (*Dermatophagoides pteronyssinus*) and the allergens it produces: identity with the house dust allergen. *Journal of Allergy*, 1967, Vol. 39, p. 325.

[30] Tovey ER, Chapman MD, Wells CW and Platts-Mills TAE. The distribution of dust mite allergen in the houses of patients with asthma. *American Review of Respiratory Disease*, 1981, Vol. 124, pp. 630–635.

[31] Korsgaard J. Mite asthma and residency. *American Review of Respiratory Disease*, 1983, Vol. 128, pp. 231–235.

[32] Sporik R, Chapman MD and Platts-Mills TAE. House dust mite exposure as a cause of asthma. *Clinical Experimental Allergy*, 1992, Vol. 22, pp. 897–906.

[33] Furumizo RT. The biology and ecology of the house dust mite *Dermatophagoides farinae*, 1973, PhD Dissertation, University of California, p. 143.

[34] Kort HSM and Kneist FM. Four-year stability of Der pI in house dust under simulated domestic conditions *in vitro*. *Allergy*, 1994, Vol. 49, pp. 131–133.

[35] Keuhr J, Frischer T, Meinert R, Barth R, Forster J, Schraub S, Burbanek R and Karmaus W. Mite allergen exposure is a risk for the incidence of specific sensitisation. *Journal of Allergy and Clinical Immunology*, 1994b, Vol. 1, pp. 44–52.

[36] Burr ML, St Leger AS and Yarnell JWG. Wheezing, dampness and coal fires. *Community Medicine*, 1981, Vol. 3, pp. 203–209.

[37] Burr ML, Miskelly FG, Butland BK, Merrett TG and Vaughan-Williams E. Environmental factors and symptoms in infants at high risk of allergy. *Journal of Epidemiology and Community Health*, 1989, Vol. 108, pp. 99–101.

[38] Martin CJ, Platt SD and Hunt SM. Housing conditions and ill-health. *BMJ*, 1987, Vol. 294, pp. 1125–1127.

[39] Platt SD, Martin CJ, Hunt SM and Lewis CW. Damp housing, mould growth, and symptomatic health state. *BMJ*, 1989, Vol. 298, pp. 1673–1678.

[40] Brunekreef B, Dockery DW, Speizer FE, Ware JH, Spengler JD and Ferris BG. Home Dampness and respiratory morbidity in children. *American Review of Respiratory Disease*, 1989, Vol. 140, pp. 1363–1367.

[41] Dekker C, Dales R, Bartlett S, Brunekeef B and Zwanenburg H. Childhood asthma and the indoor environment. *Chest*, 1991, Vol. 100, pp. 922–926.

[42] Strachan DP. Damp housing and asthma: validation of reporting of symptoms. *BMJ*, 1988, Vol. 297, pp. 1223–1226.

[43] Strachan DP and Sanders CH. Damp housing and childhood asthma: respiratory effects of indoor temperature and relative humidity. *Journal of Epidemiology and Community Health*, 1989, Vol. 43, pp. 7–14.

[44] Williamson IJ, Martin CJ, McGill G, Monie RDH and Fennerty AG. Damp housing and asthma: a case-control study. *Thorax*, 1997, Vol. 52, pp. 229–234.

[45] Scottish Homes. Scottish House Condition Survey, 1996, Scottish Homes, 1997, Edinburgh.

[46] Gotzsche PC, Hammarquist C and Burr M. House dust mite control measures in the management of asthma: meta-analysis. *BMJ*, 1998, Vol. 317, pp. 1105–1109.

[47] Walshaw MJ and Evans CC. Allergen Avoidance in house dust mite sensitive adult asthma. *Quarterly Journal of Medicine*, 1986, Vol. 58, pp. 199–215.

[48] Dorward AJ, Colloff MJ, MacKay NS, McSharry C and Thomson NC. Effect of house dust mite avoidance measures on adult atopic asthma. *Thorax*, 1988, Vol. 43, pp. 98–102.

[49] Ehnert B, Lau-Schadendorf S, Weber A, Buettner P, Schou C and Wahn U. Reducing domestic exposure to dust mite allergens reduces bronchial hyper-reactivity in sensitive children with asthma. *Journal of Allergy and Clinical Immunology*, 1992, Vol. 90, pp. 135–138.

[50] Carswell F, Birmingham K, Oliver J, Crewes A and Weeks J. The respiratory effects of reduction of mite allergen in the bedrooms of asthmatic children – a double-blind controlled trial. *Clinical Experimental Allergy*, 1996, Vol. 26 (suppl. 4), pp. 386–396.

[51] Warner JA, Marchant JL and Warner JO. Double blind trial of ionisers on children with asthma sensitive to the house dust mite. *Thorax*, 1993, Vol. 48, pp. 330–333.

[52] Huss K, Squire EN JR, Carpenter GB, Smith LJ, Huss RW, Salata K, Salerno M, Agostinelli D and Hershey J. Effective education of adults with asthma who are allergic to dust mites. *Journal of Allergy and Clinical Immunology*, 1992, Vol. 89 (suppl. 4), pp. 836–843.

[53] Charpin D, Birnbaum J, Haddi E, Genard G, Lanteaume A, Toumi M, Faraj F, Van Der Brempt X and Vervloet D. Altitude and allergy to house dust mites. *American Review of Respiratory Disease*, 1991, Vol. 143, pp. 983–986.

[54] Platts-Mills TAE and Mitchell EB. House dust avoidance. *The Lancet*, 1982, Vol. 2, pp. 1334–1336.

Chapter 4

Housing and health

There has been relatively little medical research into the domestic environment. A recent publication by the British Medical Association (BMA) entitled 'Housing and health: building for the future'[1] reviewed the available literature and concluded that housing quality is an important determinant of health. The report called for the formation of a 'Healthy Housing Taskforce' to provide a multi-sectorial approach to improving housing conditions across the UK. Given the history of slum clearance and the advent of improved sanitation and building standards throughout the twentieth century, their conclusions and recommendations may surprise many.

The UK has between 30 000 and 60 000 more deaths between December and March than in comparable four-month periods.[2] This increase is much greater than in other countries with similar or more severe climates, implying that it is not outdoor exposure to cold that is the key determinant in excess winter mortality. Northern Finland – where winter temperatures regularly drop to −20 °C – has a significantly lower rate of excess winter deaths[3] than London. Finnish dwellings have historically had much higher levels of insulation and whole-house central heating is commonplace. These additional winter deaths are mainly in the elderly population, and about 90% are registered under three generic disease headings: ischaemic heart, cerebrovascular and respiratory.[4] The biological mechanisms resulting from a lowering in core body temperatures are well known.[5] The body's defence against cold is to shut down blood vessels in the skin to reduce heat loss from the core. This displaces around a litre of blood and overloads the central organs. In order to reduce this excess, salt and water are excreted. This in turn requires more salt and water to leave the bloodstream through the walls of the blood capillaries. This adjusts the blood volume to the reduced capacity of the circulation system, but leaves the blood more concentrated. Some of the smaller molecules of the blood plasma – including the anti-thrombotic vitamin C – are able to redistribute through the capillary walls but the red and white blood cells, platelets, fibrinogen and cholesterol are too large and remain in increased concentration in the blood plasma. All promote viscosity and hypercoagulability; and an increase in the blood pressure.

Cold stress thus stimulates a range of biological processes that result in the blood becoming thicker, increasing the likelihood of cardiovascular or cerebrovascular incidents.

The immune system is also suppressed increasing the likelihood of airborne infectivity.[6] Reasons for the increase in respiratory infections are not fully understood but it appears that colder air induces bronchoconstriction and suppresses muco-ciliary defences, resulting in local inflammation. Cold air *per se* is not likely to result in respiratory infections in the absence of pathogens, as shown by a study carried out in ice-bound Spitzbergen – a town which lies inside the arctic circle – by Tyrrell.[7] Despite exceptionally cold winter air temperatures, no increase in respiratory infections occurred until the arrival of the first ship in Spring, exposing an isolated indigenous population – with possibly limited natural immunity – to urban infections.

Although some of the additional winter deaths have been ascribed by the Eurowinter Group[8] as being due to external exposure – exacerbated by inappropriate clothing levels or culturally determined behaviour – there remains an acceptance that the majority are essentially preventable, if the elderly are kept warm in their homes during the winter months. In 1997, Donaldson and Keatinge[9] demonstrated that excess winter deaths in the southeast of England from ischaemic heart, cerebrovascular and respiratory disease, had halved between the years 1977 and 1994 (Figure 4.1).

Some of this reduction can be put down to increasing housing standards and indoor temperatures resulting from major repair and improvement programmes as well as more energy-efficient new-build dwellings, confirming

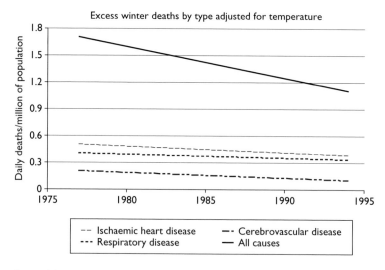

Figure 4.1 Excess winter deaths by type 1975–1995 (SE England).

the important role that house conditions play in determining public health
and the quality of life.

The impact poor housing has on health was first highlighted in 1842
by Chadwick[10] who established a clear link between poor housing conditions and a range of diseases and high mortality rates. Over 150 years
later Marsh *et al.*[11] claimed that contemporary multiple housing deprivation
appears to pose a risk to health, equal to that of smoking and greater than
that of excessive alcohol consumption. Despite such claims, housing and
house conditions presently appear to have a low profile and are well
down the political agenda, but they may be as influential in determining
public health as the nineteenth century slums that they were designed to
replace. Although there has been some reduction in excess winter deaths
since 1977 associated with cold stress, deaths from acute asthma attacks
did not follow these trends. Figure 4.2 shows deaths from acute asthma
incidents during this period and although better diagnosis and medication
appear to have halted the sharp increase, General Practitioners (GPs)
report an even greater increase in numbers presenting with symptoms[12]
(Figure 4.3).

This evidence points to a paradox. Increasing internal temperatures are
reducing some types of respiratory disease, while driving a remarkable increase
in the prevalence of asthma. If these comfort gains have been won by reducing
ventilation rates, as part of the drive for energy efficiency, dwellings will be
'tighter'. This will in turn reduce water vapour dissipation rates and increase

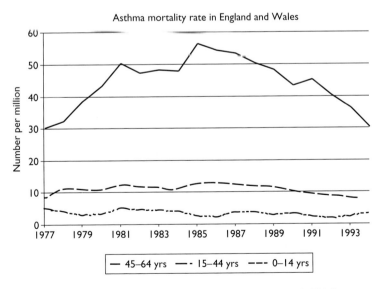

Figure 4.2 Asthma mortality rate in England and Wales (1977–1994).

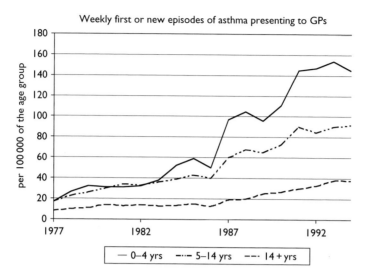

Figure 4.3 Weekly first or new episodes of asthma presenting to GPs (1977–1993).

indoor RH. There is also the potential for hazardous indoor pollutants to become more concentrated.

Although UK building regulations have seen a number of revisions over the last 25 years, there are over 20 million households in the UK and at the current rate of replacement (160 000 per annum[13]) it will take many decades for such a low 'turnover' to make a significant impact. In terms of poor energy efficiency, it is the thermal inefficiencies of the pre-1980s stock that require to be addressed with comprehensive improvement and repair programmes.

It is somewhat ironic that the increase in the prevalence of asthma appears to be as a direct result of the thrust for energy efficiency driven by the OPEC oil embargo in the mid-seventies. Energy was thrust to the forefront of the political agenda and the population faced a number of government initiatives imploring them to 'save it' or 'seal it' or 'switch it off!' The Home Energy Efficiency Scheme (HEES) has its origins in this era, with grants initially being provided for the draught-stripping of doors and windows and the installation of insulation at ceiling level in roofspaces. Such a modest and piecemeal approach was of course unlikely to be effective and the publication of the 1996 English House Condition Survey[14] demonstrated that 1 522 000 dwellings remained classified as unfit. Somewhat surprisingly this figure is greater than that contained in the 1991 survey, with dampness the prime reason for a dwelling failing the standard. Scotland, with a more severe climate, fared worse. In 1985 the first Glasgow House Condition Survey[15] identified 62 400 dwellings (38% of the total stock within the city boundary) suffering from some form of dampness, with condensation the most prevalent. The first national House Condition Survey[16] confirmed that Glasgow's problems were not unique, with 584 000 Scottish dwellings showing evidence of dampness, condensation or mould, either singly or in combination. The post-war dwelling stock exhibited the most acute problems with condensation dampness identified in 23.7% of the dwellings surveyed. Five years later, after approximately £5 billion of investment in the stock, the 1996 House Condition Survey[17] reported 'no substantive change' in the dampness statistics. It is thus possible to infer a combination of one of the following:

- The rate of housing stock deterioration in Scotland is approximately £1 billion per annum.
- The investment measures did not address the problem of dampness.
- The remedial action has exacerbated the problem.

Two questions can now be legitimately posed:

1 What have been the effects on occupant health and well-being of living in such a damp domestic environment?
2 What can be done to ameliorate or resolve these problems?

The Scottish House Condition Survey 2002[18] has subsequently reported a drop in the number of dwellings affected by dampness and/or condensation (360 000). Mould, however, was evident in 233 000 (11% of the stock) and the report demonstrated an association in children between respiratory problems and condensation (Figure 4.4).

Where dampness was reported there was a greater chance of at least one member of the household suffering from respiratory problems. The housing stock constructed after 1975, has a much lower incidence of condensation (2–7% as opposed to 12–13% in the pre-1975 stock). The survey also reported that the total cost of making good all identified disrepair was £6.5 billion. This is in addition to the estimated spend (home-owners and landlords) of £3.3 billion for the year prior to survey.

Is history repeating itself?

Throughout the nineteenth and twentieth centuries a succession of legislative measures have attempted to address issues of residential building performance and the possible effects on occupants' health.

The dramatic rise of Glasgow and its conurbation (from Greenock in the west to Wishaw in the east) over the last two centuries often resulted in overcrowding and slum conditions, as migrants from the highlands and Ireland poured into the area. Housing, at the outset of the industrial

Figure 4.4 Mould growth on damp bedroom walls.

revolution, was entirely in the hands of private sector builders and factory owners who saw it as an additional running cost, reducing the competitiveness of their enterprises. In the words of JL& B Hammond,[19] 'The quantity and quality of working class housing was decided by the avarice of the jerry builder catering for the avarice of the Capitalist.'

The outcome of this was a serious deterioration in public health with a variety of epidemics periodically decimating the migrant communities. Both central and local government felt obliged to intervene. The 1866 Glasgow Improvement Act[20] was one of the first pieces of legislation to acknowledge an unequivocal link between housing and health. It was followed by the 1875 Public Health Act[21] that attempted to improve conditions by giving local authorities specific powers to declare houses unfit and have them demolished. Although this succeeded in ensuring new houses conformed to minimum standards, Horsey,[22] has argued that it led directly to increased building costs which in turn pushed rents above a level which most workers could afford. The result was even greater overcrowding.

In 1915, despite the sacrifice of millions in northern France and Belgium, British landlords saw fit to increase rent levels. This resulted in a series of rent strikes starting on Clydeside, which quickly spread throughout the country. Central government intervened and froze rent levels for the duration of the war. This legislation was not immediately revoked after the war and directly contributed to the collapse of the private sector house building market. Finance capital will always chase high and/or secure returns. As house building and letting did not produce a competitive short-term yield, *laissez faire* economics thus proved structurally incapable of meeting the demand or improving housing conditions for the mass of people; a theme that recurs in almost every epoch.

State intervention

The Housing and Town Planning Act (1919)[21] – also known as the Addison Act – allowed local authorities to borrow money and start building and improving houses for rent. Chris Addison was a doctor who conducted surveys of the building stock and recognised the central role housing played in public health. Efforts were also being made to revive the private sector. The so-called Chamberlain Act (1923)[23] reinforced the fact that local authority support was to be considered a short-term palliative, as it provided financial aid to underwrite mortgage funds to stimulate the private market. Such limited funding resulted in the building of smaller and cheaper dwellings that were below the standards recommended by the government's advisory group (Tudors Walters Committee) in 1918. This encapsulates the crux of the matter. A well-researched report lays down clear specifications and building standards. The private sector, however, could not build to this standard and achieve a competitive return against other forms of capital accumulation. The government is then compelled to intervene with subsidy and the historical and ongoing tension between private and state capital, is realised.

The Wheatley Act[24] passed in 1924 was driven by a clear acknowledgement that long-term investment was needed to solve the housing problem with local authorities established as the main house providers. In Glasgow a total of 50 800 houses were built in the inter-war years, 35 800 being built using public funds.[25] In response to a series of misfired attempts at clearing and improving 'slum' areas, central government set up the Scottish Special Housing Association (SSHA) in the 1930s to take control of housing from the Local Authorities, aiming to stimulate employment and experiment with new building materials.[20] The Housing (Scotland) Act 1935 attempted to reduce overcrowding by specifying the maximum number of people in a house; a number which was soon to be used as a minimum standard.[26]

'Planning our new homes'

In 1942 the Scottish Housing and Advisory Committee (SHAC) was commissioned by the Secretary of State for Scotland to write a report making recommendations for the design, interior planning, layout and standard of construction of new houses in Scotland. An extensive survey was used to establish both public and professional opinions on various housing issues. In total 156 634 questionnaires were completed. The report[27] was conceived to provide guidance to Local Authority housing departments on how to build new homes, and advised on minimum standards to be achieved in each home. It was a carefully researched body of work that produced prescriptive recommendations aimed at providing warm, dry and commodious dwellings. There was also a remarkable level of agreement between public and professional opinion as to how we should build dwellings in the post-war era. Recommendations were given for the construction of all future council houses, aiming to raise the quality. These included: use of new and developing insulating materials; enhancing thermal performance; eliminating thermal isolation by employing central heating; sound proofing; weather proofing; investigations into alternative materials prior to adoption; building orientation used to best advantage; through draughts eliminated by providing independent access to the lobby; running water and private conveniences in every home.

By the mid-fifties local authorities in conjunction with central government had embarked on the largest state-sponsored house-building programme ever conceived. Housing was on the political agenda and each party's manifesto contained ambitious targets. Even a conservative government, which was returned to office in 1951, succeeded in meeting their promised quota of 300,000 houses per annum by 1953.[23]

Scottish public-sector house building peaked in 1970, with a total of 34 906 units completed. This compared with a relatively modest 8 220 private sector completions.[21] Glasgow City Council alone controlled 175 046 dwellings.[28] In 60 years public housing in Britain as a whole had increased from 1% of the overall housing market, to a staggering 64%.[22] Many of the post-war dwellings had, however, been constructed quickly and to lower standards, than the 1944 SHAC report had recommended, yet it took until 1985 to assess the stock, and the subsequent Scottish House Condition Survey – published in 1991 – was long overdue in acknowledging the extent of the dampness problems, primarily caused by low standards of insulation and a lack of energy efficiency.

Dampness and health status

In 1989 a significant association between dampness, mould growth and health status was demonstrated by Hunt *et al.*[29] This research fulfilled some of the classic criteria used to establish causality, namely, a dose–response relationship and biological plausibility. Published in the British Medical Journal the report claimed that those living in damp houses had a much greater risk of ill health, with diseases of the upper respiratory tract showing a particularly high correlation. In the Victorian era, poor sanitation and overcrowding had led to episodic epidemics of diseases such as cholera, tuberculosis and dysentery. Although such diseases are now relatively rare, changes to the design and construction of the post-war housing stock appear to be implicated in causal relationships with a range of health concerns and specific symptoms, some of which currently appear to be on the increase.

What has changed in the domestic environment?

It has been claimed that children born and bred during the war years have proven to be a remarkably robust and healthy generation. Despite the privations and stresses caused by such a traumatic socio-economic background, rationing ensured that a large number of children had relatively healthy diets. The hypothesis is simple. If you provide people with clean water, good air quality and a healthy diet, major improvements in public health can accrue. Despite rising incomes and general affluence, there are three conditions that appear to be on the increase and are particularly prevalent in west central Scotland: cancer, heart disease and asthma. The latter is intimately connected with the domestic environment and in particular hygro-thermal conditions and air quality. It is clear that the domestic environment has seen major changes during the latter part of the twentieth century, particularly when modern dwellings are compared with their nineteenth-century counterparts, which in Scotland were notoriously draughty with flues and open fires in each room driving high air change rates. Post-war construction, and more recently, comprehensive improvement and repair programmes have produced some fundamental changes which are implicated in the current asthma pandemic. The following is a list of physical characteristics and lifestyle changes that appear to be influential drivers, contributing to the increasing prevalence of the disease.

Internal volumes

The publication of the Tenant's Handbook in 1983[30] detailed the evolution of Scottish house typology over the last 100 years. Georgian and Victorian tenemental construction typically had ceiling heights between three and four metres. The abolition of part Q of the Building Regulations in 1981, effectively ended minimum space standards, as first laid down in 1918 by the Tudor Walters Committee; which were in turn revised in 1944 by the Dudley Committee and again in 1961 by Parker Morris. Modern double bedrooms at c. $12\,m^2$ can now be less than half the volume of their nineteenth-century counterparts. Although revisions to Building Regulations did ensure that energy-efficiency concerns started to be addressed in the early eighties, a study of private sector dwellings constructed by Wimpey, undertaken by Fielding,[31] demonstrated that a typical three-bedroom private dwelling had a total floor space of $60.77\,m^2$; over 20% smaller than the recommended minimum floor area laid down by Parker Morris some 40 years previously.

In 1994 Karn and Sheridan[32] published a study of space standards in 15 European countries comparing useful floor space in dwellings constructed between 1980 and 1991. The UK was ranked fifteenth, and was one of only four countries where the older housing stock exceeded the spatial standards of the new provision. A further study published in 1992 by Walentowicz[33] investigating the influence of revised cost yardsticks introduced in 1988, found that mean floor areas – for new-build housing association funded projects – had fallen by 10% since 1987/88. Even if ventilation rates had remained constant throughout this period, smaller internal volumes will result in there being less internal air to dilute and dissipate gasses, microbes, moulds, toxins and particulate matter.

Construction materials and techniques

Construction materials have different moisture capacities and absorption/ desorption coefficients. This action has been referred to as the 'sponge effect'[21] and it occurs in response to the surrounding environmental conditions, driven mainly by air temperature, moisture generation and ventilation rates.

There have been significant changes in both construction techniques and materials during the twentieth century. The Victorian tenement's walls were typically constructed of sandstone with a random rubble inner skin. Internally, the walls were finished with horse hair plaster on lath. This finish is relatively vapour permeable and can absorb and desorb water vapour with ease. The move to gypsum plaster, applied 'on hard' as well as dense concrete panel systems, significantly increased the vapour resistivity of the boundary surfaces. From the CIBSE Guide Volume A,[34] this can be estimated as up to a tenfold increase, from 20 Ns/kg to $c.$ 50–200 Ns/kg. The use of polythene vapour barriers and foil backed plasterboard, further increased the vapour resistivity of the boundary layer to between 350 and 4000 Ns/kg producing – what is in effect – an impermeable surface. In such tight, dwellings internal vapour pressure will be highly reliant on Ventilation rates to equalize to ambient levels. Older dwellings were relatively heavyweight, having solid brick internal walls and a brick inner leaf. Modern timber-frame dwellings have little thermal capacitance. The timber frame and insulation quilt isolate the brick outer leaf and lightweight stud partitions result in a fast responding internal environment, ideal for intermittent occupation cycles. The downside is an increase in diurnal temperature fluctuations when the heating system is switched off. Work by Howieson[35] confirmed internal air temperature falls of up to 14 °C within the first hour of the heating system shutdown, and a total overnight decay of almost 20 °C (average external ambient air temperature for a typical 10-hr winter period was 1.07 °C). Such a drop in air temperature will produce proportional increases in RH, with dew point likely to be reached on many surfaces. The resultant condensation can be absorbed by hygroscopic soft furnishings and carpets as well as appearing on windows and walls. The micro-climates will thus be regularly wetted – a key precursor for HDM colony establishment and proliferation.

Background ventilation

Major changes in ventilation regimes were driven by the 'Clean Air Acts' which came into being during the late 1950s and early 1960s. Flues and open fires – which until this period were common in most dwellings – started to be sealed. These effective stack ventilators – even without open fires burning – will have facilitated relatively high background ventilation and vapour dissipation rates. The introduction of low cost natural gas from the North Sea, also drove the installation of central heating systems and flued radiant appliances in urban areas, allowing much higher internal temperatures to be achieved across the dwelling. Warm air systems are a particularly effective method of transferring moisture throughout a dwelling and the associated convection currents can also disturb the dust reservoirs.

Although new dwellings now require to have the ability to be tested for air tightness (the dwelling is pressurised by a blow door fan to 50 Pa and its air leakage characteristics must fall within a given air change rate) as part of the new 'deemed to satisfy' Building Standards (Scotland) Regulations, there have been relatively few pressure tests carried out on the twentieth-century housing stock. In 1988 the Building Research Establishment (BRE) Scottish laboratory undertook a range of blow door tests on a cross-section of Scottish house types, however, the draft report circulated for comment was never published. Figure 4.5 is an extract taken from the draft, which confirms that the 1960s tenemental properties found throughout Glasgow's new estates (Castlemilk/Easterhouse/Drumchapel) were at the time exceptionally leaky, when compared with typical English house types.

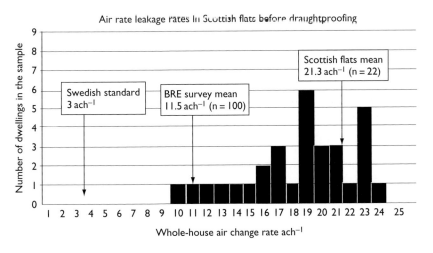

Figure 4.5 Air leakage rates in 1950s Scottish tenement flats.

Both English and Scottish examples did not compare well with the Swedish standard. It was apparently not possible to reach a 50 Pa pressure differential in some dwellings and several had to be analysed at the lower 20 Pa rate. Even when subsequently draught-stripped and re-tested, on average, only a 5% reduction was achieved. The suggestion that the construction joints were a major source of air infiltration was subsequently confirmed by thermographic photography. It is ironic that some dwellings could have few or no problems with condensation dampness because unplanned, yet fortuitously high air change rates were ensuring that water vapour builds-ups were relatively quickly dispersed, diffused and expelled.

The widespread availability of grants for draught-stripping programmes brought in during the 1980s under the government's HEES may also have reduced air change rates, with the specific aim of increasing comfort and/or reducing fuel consumption. Major modernisation and improvement programmes that were actioned in the 1980s, were to change the stocks characteristics in a more fundamental manner.

Double-glazing

Between 1991 and 1996 the percentage of dwellings in Scotland which had double-glazing increased from 36 to 62%.[17] As less than 5% of this increase was due to new-build dwellings, approximately 550 000 existing dwellings had replacement units fitted during this five-year period. This compares with only a 3% increase in wall insulation over the same time frame. In the public sector these windows were primarily PVC units which replaced the original sash and case timber casements or notoriously draughty side-hung single-glazed steel units. In dwellings where double-glazing was the sole improvement measure (i.e. no complementary upgrading of insulation or commensurate ventilation facilities such as extract fans or trickle vents etc.) it is highly probable that internal water vapour pressures would rise. Although the standard advice disseminated by public-sector landlords as reported by the Scottish Consumer Council[36] was to, 'turn up your heating and open your windows', increasing air change rates by opening windows, simply allows expensive heat to escape. The choice appears to be between draughty windows that render the dwelling 'hard to heat' and 'thermally unsafe',[37] or double-glazing, insulation and central heating, with a resultant increase in internal water vapour pressures and air temperatures, providing ideal conditions for HDM colonisation and proliferation.

A study undertaken by Revie and Howieson,[38] comparing two communities in markedly different house types in Paisley, found that residents living in modern dwellings with double-glazing and central heating had three times the prevalence of asthma than those living in older, less energy-efficient stock.

Bathing and clothes drying

In 1951 Mather and Crowther[39] published a survey for the Coal Utilisation Council which reported that only 46% of dwellings had bathrooms, 56% had hot water, and the average person took 1.4 baths per week. The 1996 Scottish House Condition Survey[17] reported that 99.6% of dwellings had baths. In addition the trend towards fitting separate or integral showers may be significant, as showers produce a relatively high water vapour aerosol effect which in turn increases evaporation rates and internal water vapour pressures.

In 1991 the BRE[40] estimated that the average moisture burden in a typical British household was between 7 and 14 litres of water vapour per day, with clothes drying alone accounting for 4 litres. About half this moisture is produced slowly throughout the day in different rooms of the home. The remainder is produced over short periods of time and in large quantities, mainly in the kitchen and bathroom. The increasing incidence of female employment and high crime rates in working-class schemes may also have driven an increase in internal clothes drying. This may not be a problem where ventilation rates are high, but when taken in combination with the range of other changes outlined, water vapour and moisture produced within a dwelling will start to accumulate, particularly during the winter months.

Soft furnishings and internal surfaces

It was common in Victorian tenements to have exposed floorboards with loose scatter rugs. There was also a dedicated profession of carpet beaters, who would tour an area and extract a household's rugs to be thoroughly beaten in the back 'green'. Such a process may have resulted in the effective dispersal of allergens and house dust to the external air. The advent of fitted carpets and vacuum cleaners may not have been such an effective solution, as it provided extensive micro-climates for dust mite colonisation and before the advent of HEPA (High-efficiency particulate air) filters small particulate matter could easily pass through the bag. The UK has the highest consumption of carpets in Western Europe and North America[41] at 3.9 m^2 per person with 98% of British homes having fitted carpets. This compares with 16% in France and 2% in Italy. Foam-based furniture and the now ubiquitous soft toy, have also increased the opportunity for suitable HDM habitats. Strachan and Carey[42] for instance, found eight times higher levels of dust mite allergens in synthetic than traditional feather pillows – which are routinely treated with chemicals to kill parasites – suggesting that soft toys may well be a significant HDM habitat.

Warm and humid indoor environments

As tenure has swung back in favour of the private sector during the latter part of the twentieth century, there have been some fundamental changes in the design, construction and patterns of use in domestic buildings. Modern dwellings have become smaller and 'tighter' and these changes when combined with the significant alterations in lifestyle, such as daily showering and central heating, result in the creation of warmer and more humid indoor conditions. The use of materials with a higher vapour resistance and/or the lack of thermal capacitance in timber-frame dwellings will have resulted in much higher water vapour pressures – leading to diurnal condensation cycles – that can produce the ideal micro-climates in carpets, bedding and soft furnishings for the colonisation and proliferation of the HDM.

The housing stock has become more energy efficient due in part to revisions in new-build standards combined with comprehensive modernisation and improvement programmes, retrofitting the existing stock. Over the period 1996–2003 the energy efficiency of the Scottish housing stock increased by *c.* 4% (4.1–4.5 using the National Home Energy Rating index (NHER) and from 43 to 46.7 using the Standard Assessment Procedure (SAP) rating).[43] The net effect is to make dwellings marginally less difficult to heat and – if expenditure on fuel remains constant – more warmth can be purchased, leading to increased internal temperatures. A pilot study[44] into the effect of the HEES – providing grants for low income groups to underwrite draught-stripping, loft and cavity insulation – confirmed a modest increase

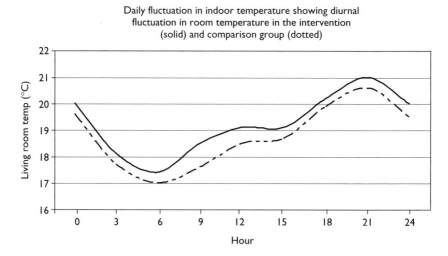

Daily fluctuation in indoor temperature showing diurnal fluctuation in room temperature in the intervention (solid) and comparison group (dotted)

Figure 4.6 Daily fluctuation in indoor temperatures before and after HEES improvements.

in both bedroom and living room temperatures (Figure 4.6). The report however concluded,

> The study found a number of features militating against health improvements. These include twice the level of mould growth found on average in the English housing stock; a poor level of ventilation (3 out of 10 homes with no purpose provided ventilation) and almost all had conditions favourable to house dust mites. There were no appreciable signs of the intervention works having improved overall conditions or specifically reduced the health risks associated with dust mites.

The evidence base supporting the primary hypothesis, that it is HDM allergens that are driving the asthma pandemic in the UK, appears to be growing.

References

[1] British Medical Association, Housing and health: building for the future, London, May 2003.

[2] Boardman B. Fuel poverty: from cold homes to affordable warmth, Belhaven Press, London, 1991.

[3] Eurowinter group, Cold exposure and winter mortality from ischaemic heart disease, cerebrovascular disease, respiratory disease and all causes in warm and cold regions of Europe. *The Lancet*, Issue 9062, May 1997, pp. 1341–1346

[4] Curwen M. Excess winter mortality in England and Wales with special reference to the effects of temperature and influenza, The Health of Adult Britain, 1997. Vol. I and II (Dicentennial Supplement), No. 12, The Stationary Office, London.

[5] Keatinge WR, Coleshaw SRK, Cotter F, Mattock M, Murphy M and Chelliah R. Increase in platelet and red cell counts, blood viscosity and arterial pressure during mild surface cooling: factors in mortality from coronary and cerebral thrombosis in winter, *BMJ*, 1984, Vol. 289, London, pp. 1405–1408.

[6] Collins K. Cold, cold housing and respiratory illness, Cutting the Cost of Cold: Affordable Warmth for Healthier Homes, Chapter 3 (eds J Rudge and F Nicol), London, 2000.

[7] Tyrrell DAJ. *Common Colds and Related Diseases*, Edward Arnold, London, 1965.

[8] Eurowinter Group, op. cit.,

[9] Donaldson GC and Keatinge WR. Mortality related to cold weather in elderly people in South east England 1970–1994, *BMJ*, 1997, Vol. 315, pp. 1055–1056.

[10] Chadwick E. Report on the sanitary conditions of the labouring population of Great Britain, HMSO, London, 1842.

[11] Marsh A, Gordon D and Pantazis C *et al*. Home sweet home? The impact of poor housing on health, The Policy Press, Bristol, 1999.

[12] Royal college of general Practitioners Birmingham Research Unit, Weekly Returns Service, England and Wales 1976–2000 in Out in the Open. A true

picture of asthma in the UK today, National Asthma Campaign, September 2001, Vol. 6, No. 3.

[13] Department of Trade and Industry, Fuel Poverty Strategy, 2001 The Stationary Office, London.

[14] Department of the Environment, Transport and Regions, English House Condition Survey, HMSO, London, 2000.

[15] House Condition Survey 1985, The Condition of Glasgow's Housing, Vol. 1.

[16] Scottish Homes. Scottish House Condition Survey 1991, Scottish Homes, Edinburgh, 1991.

[17] Scottish Homes. Scottish House Condition Survey 1996, Scottish Homes, Edinburgh, 1997.

[18] Communities Scotland. Scottish House Condition Survey 2002, Edinburgh 2003, ISBN 1 874170 54 1, Chapter 1, p. 13.

[19] Hammond JL&B. The Town Labourer 1760–1832, Kelly, London, re-printed 1917.

[20] Clapham D, Kemp P and Smith SJ. *Housing and Social Policy*, Macmillan Education Ltd, London, 1990.

[21] Begg T. Housing Policy in Scotland, John Donald Publishers Ltd, Edinburgh, 1996.

[22] Horsey M. Tenements and Towers, Glasgow Working Class Housing 1890–1990, The Royal Commission on the Ancient and Historical Monuments of Scotland, Edinburgh, 1990.

[23] Balchin PN. *Housing Policy and Housing Needs*, Macmillan Press Ltd, London, 1981.

[24] Nuttgens P. *The Home Front, Housing the People from 1840–1990*, BBC Books, London, 1989.

[25] Burnett J. *A Social History of Housing*, The University Press, Cambridge, 1975.

[26] Crammond RD and Sir John Mactaggart. Housing Policy in Scotland 1919–1964, A Study in State Assistance, Oliver & Boyd, 1996, London and Edinburgh.

[27] The Scottish Housing Advisory Committee, Planning Our New Homes, HMSO, 1944.

[28] House Condition Survey 1985. The Condition of Glasgow's Housing, Vol. 1, Glasgow City Council, Glasgow, 1986.

[29] Hunt SM, Martin CJ, Platt SD, Lewis C and Morris G. Damp housing, mould growth and health Status, Part I. *British Medical Journal*, 1989, Vol. 298, p. 1673.

[30] Gilbert J, Orr C and Hashagen S. The Tenant's Handbook: A guide to the repair maintenance and modernisation of Council Houses in Scotland, Assist Architects, Scottish Consumer Council, Glasgow, 1983.

[31] Fielding N. The volume housebuilders, Roof, November/December 1982, pp. 16–18.

[32] Karn V and Sheridan L. This small world is no wonderland. Inside Housing, 21 July 1994, pp. 12–13.

[33] Walentowicz P. Housing standards after the Act. A survey of space and design standards on housing association projects in 1989/90. Research report 15, NFHA Research, 1992.

[34] CIBSE Guide, Volume A, Design data. The Chartered Institution of Building Services Engineers, 5th Edition, London, 1974.

[35] Howieson SG. *Housing-Raising the Scottish Standard*, Technical Services Agency, Glasgow, 1991.

[36] Scottish Consumer Council. Houses to mend – A survey of council house repairs in Scotland, Glasgow, 1978.

[37] Howieson SG. *Housing: Rasing the Scottish Standard*, Technical Services Agency, Glasgow, 1991.

[38] Revie C and Howieson SG, Fuel Poverty and Health in Paisley, Energy Action Scotland, Glasgow, 1999, p. 154.

[39] Mather and Crowther, Survey for Coal Utilisation Council, London, 1951.

[40] Garratt J and Nowak F. Tackling condensation, Building Research Establishment, Garston, Watford, 1991.

[41] Warner JA, Little SA, Pollock I *et al.* The influence of exposure to house dust mite, cat, pollen and fungal allergens in the home on primary sensitisation in asthma, *Paediatric Allergy and Immunology*, 1990, Vol. 1, pp. 79–86.

[42] Strachan D and Carey I. Reduced risk of wheezing in children using feather pillows is confirmed, *BMJ*, 1997, Vol. 314, p. 518.

[43] Scottish Homes/Communities Scotland, Scottish Housing Condition Surveys 1996 & 2002, Edinburgh.

[44] UCL, Report on the pilot health impact assessment of the new home energy efficiency scheme, Centre for Regional and Economic Social Research, London, 2001.

Chapter 5

Historical changes in domestic ventilation regimes

Modelling five generic house types

There is relatively little research into ventilation rates in twentieth-century dwellings, and due to major modernisation and alteration programmes, it is difficult to identify a statistically significant subset of dwellings in their original configuration. The approach left open is to compare possible changes over the period by constructing computer-based models and running theoretical simulations, which can at least norm reference the likely scale of any changes. Two key questions have to be addressed: (i) What is the likely scale of any reduction in domestic ventilation rates? (ii) What impact will any reduction in air change rates have on water vapour burdens and RH?

ESP-r[1] is an integrated modelling tool developed by the Energy Simulation Research Unit (ESRU) at the University of Strathclyde to investigate thermal, visual and acoustic performance of buildings. Five traditional Scottish house-types were modelled to norm reference air change rates in a typical living room for a 48-hr winter period. By inputting a notional amount of water vapour into the model, a mass flow calculation can then estimate the rate of moisture diffusion and/or accumulation.

The 1991 Scottish House Condition Survey[2] defined housing in Scotland by type, style and construction date – each representing the respective percentage of the total housing stock (2 232 000 units). The following have been identified by Gilbert, Orr and Hashagen[3] as the five most common generic types associated with twentieth-century house-building epochs in the Glasgow area (as shown in Figure 5.1).

The main construction characteristics (volume, materials, heating system, flues, window type, crack length and trickle vents) were input as boundary conditions for a typical living room volume (largest domestic volume/occupancy rate). Simulations were then run to estimate air change rates during a 48-hr mid-winter period (Table 5.1 – Jan 21st–22nd – composite climate file).

The most apparent difference between the two earlier and the three later models is the impact of the 'Clean Air Acts' on the presence and use of open

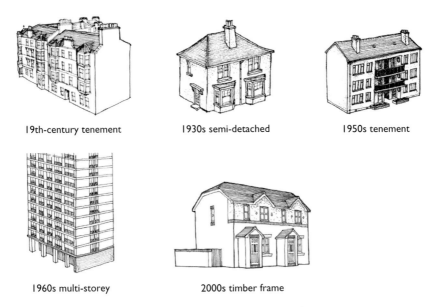

Figure 5.1 Generic Scottish house types typical of the epoch.

Table 5.1 Ach^{-1} for five living room models

	Max. (ach^{-1})	Min. (ach^{-1})	Ave. (ach^{-1})	Vol. (m^3)	(m^3/h)
1890 model	2.15	0.78	1.66	83	138
1936 model	2.00	1.10	1.63	42	69
1950 model	1.03	0.63	0.83	37	31
1970 model	0.95	0.05	0.74	41	31
2000 model	0.63	0.23	0.45	33	15

flues. The air change rates for the 1890 and 1936 models at the chimney node (ESP-r calculates flow rates at a variety of pre-determined node points) within the airflow network were 12 ach^{-1} and 10 ach^{-1} respectively, providing evidence of the impact a flue has on the total room air change rate. An open fire would drive even greater air change rates, however, it was not possible to model the variable dynamics of a coal-fired combustion process.

In terms of volumetric airflow, the late Victorian tenement has over nine times the rate of a contemporary timber-frame model. Although the simulations are not necessarily attempting to criterion reference the models to produce an accurate facsimile of reality, the method can be defended as a technique to benchmark longitudinal trends and identify the influential factors (wind speed and direction) driving the system. Both the ASHRAE and CIBSE[4] minimum recommendations of 8–10 l/s per person for odour control equate to a supply volume, for one individual, of between 29 and 36 m^3 per hour – almost twice the background air change rate of the twenty-first-century model. The ASHRAE requirement for smoking lounges is 30 l/s per person, which equates to a mass flow rate approaching 110 m^3/hr (c. 3 ach^{-1} for each smoker occupying a modern living room). Such low rates in tight modern dwellings – although in line with those implicit in the UK Building Regulations (4–5 l/s) – will allow odour, smoke and a multiplicity of hazardous indoor pollutants to build during the winter when windows are likely to remain closed. Multiple occupation will of course compound these problems. The implications for vapour dissipation rates are of equal concern.

Moisture dissipation rates

Ventilation rates have an important impact on internal water vapour diffusion. The following calculations are based on the mass flow rate of moisture, relative to the specific air change rates for each generic house type. It is assumed that the external air entering the living room through one inlet is mixing perfectly with the 'moist' air in the zone. The calculation is based solely on the living room as the test bed, with notional input and output boundary conditions for a typical winter day (incoming air temperature 4 °C, relative humidity 80%, mixing ratio 4 g kg^{-1}, specific volume 0.7898 m^3 kg^{-1}). The internal conditions are determined using the resultant data from the simulations: temperature 21 °C, air change rate, volume and flow rate specific to each model. The outgoing air is set to a temperature of 21 °C and an RH of 70% (as this is the boundary condition for mould growth[5]) and 50% RH, a figure well below known HDM viability.[6]

The volume flow rate is calculated for each model using the volume of the room and the associated air change rate. This is then used to calculate the mass flow rate of air, which when multiplied by the moisture difference between the internal and external air provides the mass flow rate of moisture. Moisture dissipation rates can then be determined assuming that all the moisture leaves the room. Three litres of water vapour was introduced into the model that equates to a pro rata (m^2) proportion of the average daily household production[5] (Figure 5.2).

Despite the relatively low mixing ratio of the external air input, the 2000 model dwelling, is produced to take almost 40 hr to reach the 70% RH boundary. It is thus more likely that relatively tight, energy-efficient, modern dwellings will be subject to progressive and cumulative moisture build-ups during the winter months if windows remain closed. They are also

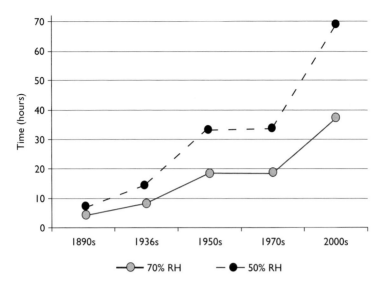

Figure 5.2 Moisture dissipation rates (hours taken to reach 70% and 50% RH boundary).

likely to have water vapour burdens that allow RH to regularly rise above HDM viability thresholds. Furthermore, diurnal temperature variations – particularly in lightweight construction that has little thermal capacitance – will be greater, increasing RH and condensation rates, which can be absorbed by carpets, bedding and soft furnishings. As ambient air in the warmer months – particularly along the western seaboard – has in the main a higher mixing ratio than $7\,\mathrm{g\,kg^{-1}}$ of dry air, there is less scope for reducing the internal absolute humidity below dust mite viability levels. Winter ventilation rates are thus an important variable in inhibiting colony size and activity, especially if a dwelling can be kept warm, suppressing internal RH.

Blow door site trials

In an attempt to partially validate the computer modelling, blow door tests were carried out on the five generic house types. The test required a high velocity fan to be fitted in a frame of the main entrance door (Figure 5.3) and the dwelling pressurised or depressurised to create a differential of $50\,\mathrm{Pa}$. Measuring the mass flow rate of air can then determine the air change rate, given a specific house volume.

Figure 5.3 Blow door fan rig.

All dwellings were randomly identified but it proved impossible to find and achieve access to both the 1930s and 1970s multi-storey dwellings, in their original condition. Both dwellings tested had new double-glazed windows fitted and all but one flue in the 1930s dwelling had been blocked.

Comparing house type performance

Type 1: Nineteenth-century four-storey tenemental (first floor, 3 apartments)

Address: Clarkston Rd, Glasgow (c. 1900)

Type and construction:

Three-apartment first-floor flat – sandstone with plaster on lath;
Single glazed timber centre pivot windows; three open flues.
Floor area: 65 m^2; Volume: 200 m^3
External wall area: 256 m^2
Calculated Air leakage rate at 50 Pa: 15.3 ach^{-1}
Mass flow rate of air at 50 Pa: 3071 m^3/hr
Air permeability @ 50 Pa: 12 m^3/hr/m^3
Effective leakage area (cm^2@ 4 Pa): 1094
Estimated background air leakage range (divisor of 10–30): 1.53–0.51 ach^{-1}
Estimated background mass flow rate of air (divisor of 10–30): 307–102 m^3/hr

Type 2: 1930s semi-detached villa

Address: Warden Road, Knightswood, Glasgow (c. 1934)

Type and construction:

Four apartment semi-detached – cavity brick with plaster
on-hard; suspended timber floors;
timber pitched and slated roof, double-glazed; one flue
Floor area: 45 m^2 × 2   Volume: 246 m^3
External wall area: 241 m^2
Calculated air leakage rate at 50 Pa: 11 ach^{-1}
Mass flow rate of air at 50 Pa: 2709 m^3/hr
Air permeability @ 50 Pa: 11.2 m^3/hr/m^2
Effective leakage area (cm^2@ 4 Pa): 515
Estimated background air leakage range (divisor of 10–30):1.1–0.36 ach^{-1}
Estimated background mass flow rate of air (divisor of 10–30): 271–90 m^3/hr

Type 3: 1960s three-storey tenemental

Address: Arnprior Quad, Castlemilk, Glasgow (c. 1958)

Type and construction:

Four apartment top-floor flat – cavity brick with plaster on hard; suspended timber
floors; timber pitched and slated roof;
single glazed steel windows; no open flues
Floor area: 65 m^2 Volume: 161 m^3
External wall area: 233 m^2
Calculated air leakage rate at 50 Pa: 20.5 ach^{-1}
Mass flow rate of Air at 50 Pa: 3314 m^3/hr
Air permeability @ 50 Pa: 14.2 m^3/hr/m^2
Effective leakage area (cm^2@ 4 Pa): 838
Estimated background air leakage range (divisor of 10–30): 2.05–0.68 ach^{-1}
Estimated background mass flow rate of air (divisor of 10–30): 331–110 m^3/hr

Type 4: 1960s multi-storey

Address: Kirkton Ave, Knightswood, Glasgow (c. 1964)

Type and construction:

Four apartment first floor flat – concrete frame with cavity brick infill
plastered on hard; concrete floors; double-glazed PVCu replacement windows;
mechanical ventilation in internal bathroom.
Floor area: 67 m²; Volume: 158 m³; External wall area: 212 m²
Calculated Air leakage rate at 50 Pa: 5.3 ach⁻¹
Mass flow rate of air at 50 Pa: 843 m³/hr
Air permeability @ 50 Pa: 4 m³/hr/m²
Effective leakage area (cm²@ 4 Pa): 78
Estimated background air leakage range (divisor of 10–30): 0.53–0.18 ach⁻¹
Estimated background mass flow rate of air (divisor of 10–30): 84–28 m³/hr

Type 5: 2000s semi-detached timber frame

Address: Oak Wynd, Cambuslang, Glasgow (2000)

Type and construction:

Four apartment semi-detached villa, timber frame with brick skin;
plasterboard linings; suspended timber floors; timber pitched and tiled roof;
double-glazed PVCu windows, open vent in kitchen
Floor area: 32 m$^2 \times$ 2; Volume: 160 m^3 External wall area: 187 m^2
Calculated Air leakage rate at 50 Pa: 12.6 ach^{-1}
Mass flow rate of air at 50 Pa: 2019 m^3/hr
Air permeability @ 50 Pa: 10.8 m^3/hr/m^2
Effective leakage area (cm^2@ 4 Pa): 451
Estimated background air leakage range (divisor of 10–30): 1.26–0.42 ach^{-1}
Estimated background mass flow rate of air (divisor of 10 30): 202 67 m^3/hr

Although only three of the dwellings can be considered to be in their 'as-built' condition, the blow door tests support the hypothesis that ventilation rates are likely to have fallen during the twentieth century and that the scale of this change could be a significant determinant in indoor air quality. Victorian tenements were constructed with relatively high ceilings and thus had a greater volume per square metre of floor area. Open flues – even without coal fires burning – facilitate relatively high mass flow rates. When the two tenemental properties (1900s and 1960s) are compared with the other three, a significant differential in mass flow rates per square metre floor area, is evident. This is less than the figures generated by the ESP-r computer modelling but it has to be borne in mind that these are whole-house rates whereas the initial ESP-r simulations limited their focus to modelling the living room in isolation. As we do not spend much time occupying hallways and corridors, air quality issues are essentially about conditions in the living and bedrooms.

Of particular concern is the multi-storey flat which had only 5.3 ach^{-1} at 50 Pa, and a mass flow rate of 843 m^3 per hour. This represents a ventilation rate of between 0.18 and 0.53 ach^{-1} and it is not surprising that the tenant complained that the advent of new double glazed windows units coincided with the emergence of mould growth and more visible condensate run-off on window sills.

If a divisor of 10 is used to convert air leakage rates at 50 Pa to background air leakage under normal climatic conditions, the results of the modelling (living room only) can be compared with the test results (whole house) if converted to mass flow rate per hour, per square metre of floor area (Table 5.2).

The modern timber-frame dwelling appears to be relatively air leaky. Most of the pressure loss was, however, accounted for by two mechanical extract vents in the kitchen and bathroom. The kitchen vent did not have external louvres fitted and thus represented a large hole in the external wall, compromising the overall integrity of the envelope. When louvres stopping back-draughts are fitted, the air infiltration rate (as opposed to the air leakage rate) is likely to reduce significantly. The 1950s flat tested using the blow door was a top-floor flat and thus the external area and the additional air leakage paths through the ceiling could account for the differential. The 1930s model had double-glazing recently retrofitted, whereas the ESPr simulation presumed original draughty timber windows.

The results of such limited testing must of course be treated with caution and a much larger subset will require to be tested to validate the computer model. The tests, however, highlight the significance of measuring air leakage rates in individual rooms rather than the dwelling as one large volume. Most time indoors will be spent in living rooms and bedrooms. It is thus important to look at ventilation rates in these smaller volumes and measure any changes that can be brought about by the retrofitting of MHRV systems. This has significance for appropriate remediation strategies and the development of metrics that can identify the significance of any changes in air quality and allergen burdens.

Table 5.2 Modelled living room vs whole-house pressure test in m^3/hr/m^2 floor area

	Ave. (ach^{-1})	(m^3/h)	(m^3/hr/m^2)		Ave. (ach^{-1})	(m^3/h)	(m^3/hr/m^2)
1890 model	1.66	138	**5.3**	test	1.53	307	**4.7**
1936 model	1.63	69	**3.9**	test	1.1	271	**3.0**
1950 model	0.83	31	**1.9**	test	2.05	331	**5.0**
1970 model	0.74	31	**1.7**	test	0.53	84	**1.5**
2000 model	0.45	15	**1.0**	test	1.26	202	**3.1**

References

[1] Clarke J. Energy Simulation in Building Design. Oxford, Butterworth-Heinemann, 2001.

[2] Scottish Homes. The 1996 Scottish House Conditions Survey, Edinburgh, 1997.

[3] Gilbert J, Orr C and Hashagen S. The Tenant's Handbook: A guide to the repair maintenance and modernisation of Council Houses in Scotland, Assist Architects, Scottish Consumer Council, Glasgow 1983.

[4] CIBSE Guide Section B2. Chartered Institution of Building Service Engineers, London, 1986. Also ASHRAE, American Society of Central Heating, Refrigeration and Air-Conditioning Engineers Ventilation Guidelines, Ventilation for Acceptable Air Quality, Standard 62 – 1989.

[5] Garratt J and Nowak F. Tackling condensation, Building Research Establishment, Watford, 1991.

[6] Colloff M, Ayres J, Carwell F, Howarth P, Merrett TG, Mitchell EB, Walshaw M, Warner JO, Warner JO and Woodcock AA. The control of allergens of house dust mites and domestic pets: a position paper. Clinical and Experimental Allergy, 1992, Vol. 22 (suppl. 2), pp. 1–28.

Chapter 6

Designing a double-blind placebo-controlled interventionist trial

Selecting the team

Any investigation into the relationship between indoor hygro-thermal conditions, dust mite allergens and asthma requires an understanding of four distinct fields – the built environment, environmental physics, the physiology of asthma and the properties and effects of HDM allergens. The preceding chapters have examined the possible role increasing indoor humidity has played in allowing HDM colonisation and proliferation and established a clear causal link between HDM allergens and the aetiology of the asthmatic state. Running throughout has been an understanding that well-chosen building interventions, which can increase ventilation rates and extract moisture burdens at source, will in turn inhibit HDM activity and allergen production.

Due to the culture of compartmentalisation prevalent in both the building professions and university faculties, many research projects are conceived from, and evolve with, a mono-disciplinary bias. This may result in the research programme emphasising one particular aspect, where a more balanced, multi-factorial approach will prove more productive. A team was assembled consisting of an architect, an environmental engineer, a public health epidemiologist, a pulmonary physician, a clinical immunologist and a statistician. A methodology was then developed to answer the following research questions: Taking a sample of randomly identified dwellings in public-sector housing estates (west central Scotland – Bellshill/Uddingston/Tannochside/North Motherwell – areas in receipt of additional poverty relief funding from North Lanarkshire Council),

1 What percentage of dwellings contain dust mite allergen (*Der pI*) reservoirs above the WHO sensitisation thresholds and therefore have the capacity to sensitise atopic individuals and produce asthmatic symptoms?
2 What are the average annual and seasonal ranges in internal RH in bedrooms, and is RH positively correlated with dust mite activity?
3 How effectively can low-cost allergen reservoir denaturing, removal and avoidance measures suppress antigen levels?

4   How effective are cartridge MHRV units in suppressing absolute and relative humidity in bedrooms?

5   Will increased ventilation rates, in combination with an allergen removal/avoidance regime, produce any measurable improvements in the lung function of diagnosed asthmatics?

6   Will such a strategy prove cost effective?

Research objectives

The following list of objectives were devised in order to deliver answers to the research questions:

1   To identify robust techniques to sample and analyse domestic dust reservoirs and quantify *Der pI* levels.

2   To design a robust longitudinal and accurate thermo-hygro monitoring regime.

3   To identify a methodology for assessing a dwelling's energy efficiency.

4   To identify a set of effective low-cost allergen avoidance/removal techniques.

5   To specify an MHRV unit capable of providing an increase in air change rates without incurring an energy cost penalty.

6   To implement a dust-sampling regime capable of monitoring *Der pI* concentrations over a 24-month period.

7   To design a regime to monitor respiratory function over a 24-month period.

8   To produce a questionnaire capable of monitoring patients' attitudes and self-assessed respiratory health over a 24-month period.

9   To identify the cost of the cohorts' asthma to society in terms of primary or acute treatment and compare this with the capital and running costs of the remedial measures.

Research methods

The trial was designed to be double blind and placebo controlled, and although the subjects were volunteers, they were randomly distributed into an active and two control groups. The main thrust of the study was to identify whether the removal of dust mite colonies and their allergens from domestic dwellings would impact on asthmatic symptoms and whether a reduction in internal winter humidity – brought about by the additional ventilation measures – would give added value to the more common denaturing and allergen avoidance regimes. The intervention thus concentrated on measures to reduce both dust mite allergen reservoirs and internal humidity. Three particular methods were eventually selected.

Steam cleaning of carpets

The use of steam cleaning was chosen in preference to freezing or the use of acaricides (poisons), as it requires little expertise to operate the equipment, is less hazardous to use, and does not involve toxic chemicals. The steamer selected was a unit developed by Medivac Health (Medivac Healthcare Ltd, Wilmslow House Grove Way, Wilmslow, Cheshire, SK9 5AG). It is reasonably portable and produces super heated steam (*c.* 120 °C) under pressure (6 bar), effectively 'cooking' the water soluble allergenic proteins contained in the dust reservoirs (Figure 6.1). Tests already completed by the manufacturers claimed an efficacy of between 80 and 90% allergen reduction. Four pilot trials were undertaken prior to finalising the protocol. Care was taken to move the steamer head over the carpet at a regular rate – (*c.* 2 m^2 per minute – approximately 10 min for a typical living room of 20 m^2). Dust sampling (before and after) confirmed the efficacy of such a treatment. The efficiency of this steam cleaning system in killing mites and denaturing the allergens entrained in the mattress was never tested. Proprietary steam 'pokers' have been developed for this purpose but require the mattress to

Figure 6.1 Steam cleaning of carpets.

be punctured – an approach that may not have found favour with the volunteer cohort. For this reason a method of avoidance rather than denaturing was selected.

New bedding

As bedding (blankets/duvets/pillows) will absorb heat and sweat from recumbent occupants for *c.* 8 hr per day, it provides a warm and damp habitat, ideal for the HDM and consequentially has been found to contain high concentrations of dust mite allergens. For this reason new pillows and duvets were supplied to the active group and one control group to reduce exposure to allergens when sleeping. Barrier bedding – manufactured from breathable micro-weave material impervious to dust mite allergens – was employed to encapsulate mattresses. The chosen barrier bedding (Medivac cotton barrier bedding range) fully encases the mattress and includes extra protection around the vulnerable zip area where allergens are occasionaly released in low quality mattress covers.

Mechanical heat recovery ventilation units

Baxi E100 units were chosen due to their low capital cost and ease of installation (Baxi Clean Air Systems Ltd, Bamber Bridge, Preston). They incorporate a cross-flow heat exchanger which pre-warms the fresh incoming air stream, while stale inside air is simultaneously extracted. Using this method, around 50–60% of heat can be recovered (Figure 6.2).

Three whole-house units were also included to provide a scoping comparison. These units extract air from the kitchen and bathroom and supply air, via a heat exchanger mounted in the roof space, to the living room and bedroom of the asthmatic patient. Such systems have the advantage of being discrete, inaudible and difficult to damage or switch off. Extracting water vapour at source is likely to be more efficient than allowing it to migrate through the dwelling. The units however require duct runs to be boxed-in and re-decoration incurs additional costs. The whole-house systems are approximately six times more expensive to install and incur a similar increase in running costs (45 W versus 15 W equates to £27 per annum versus £9 per annum per cartridge unit).

As the ambient absolute humidity in West central Scotland is generally below the $7\,g\,kg^{-1}$ of dry air (the reported HDM viability threshold) for around 65% of the year, the units extract humid internal air, which is replaced with pre-warmed 'dry' external air. Such a regime should lower the internal humidity, reducing the risk of condensation dampness and the occurrence of mould, as well as progressively drying out the HDM micro-climates in bedding, carpets and soft furnishings.

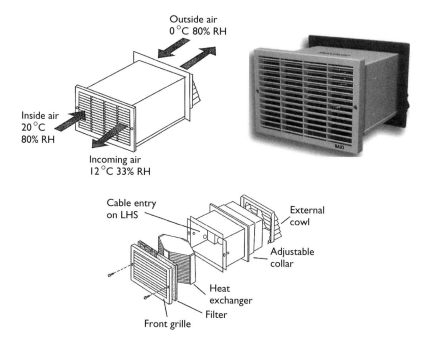

Figure 6.2 Baxi E100 reclamation unit.

New technologies have produced a significant reduction in fan noise and thus the user compatibility can be greater than with extractor fans that have typically been installed in public-sector kitchens and bathrooms. In an average public-sector bedroom ($35\,m^3$), the fans – which have a flow rate of $43\,m^3$/hr ($12\,l/s$) – will induce an increase in the air change rate of over $1\,ach^{-1}$. As they have been designed to run continuously this should result in an additional 8760 air changes per year. A boost facility with an increased flow rate is under user control and can increase the ventilation rate up to $77\,m^3$ of air per hour. Such a facility is designed for use during periods of high water vapour production or when vacuuming is undertaken to expel temporarily airborne allergen loads. It can also be used when the room has increased occupation and to extract environmental tobacco smoke (ETS).

As dust mite levels vary seasonally, it was decided that home visits would be conducted on a quarterly basis for the purpose of: downloading and restarting data loggers; collecting dust samples; recording gas and electricity consumption; completing health questionnaires and collecting completed peak flow charts which were measuring daily (morning and evening) variations in lung function.

Intervention measures

Active group 2 (all measures)

- Steam cleaning living-room and asthmatics' bedroom carpets;
- New bedding – mattress encapsulation;
- Active MHRV fans.

Active group 2 attempted to measure the specific effect MHRV units have on reducing internal water vapour pressures and any associated impact on dust mite re-colonisation rates (Figure 6.3). Three dwellings, however, had half-house MHRV units fitted where air was extracted from the kitchen and bathroom and pre-warmed air delivered to the bedroom and living room. A comparison with the individual cartridge units can thus measure the respective efficacy of each system.

Active group 1 (selective measures) (Figure 6.4)

- Steam cleaning living-room and asthmatics' bedroom carpets;
- New bedding – mattress encapsulation;
- Placebo MHRV fans.

Control group (placebo measures) (Figure 6.5)

- Placebo fans (recirculating internal air);
- Placebo steam cleaning (brush head not applied to carpet surface);
- No change to bedding.

These measures were applied after an initial six-month monitoring period to establish baseline internal hygro-thermal conditions, existing allergen

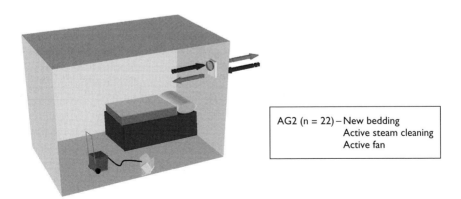

AG2 (n = 22) – New bedding
Active steam cleaning
Active fan

Figure 6.3 Active group 2 intervention measures schematic.

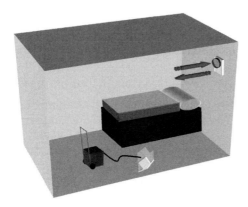

AG1 (n = 10) – New bedding
Active steam cleaning
Placebo fan

Figure 6.4 Active group I intervention measures schematic.

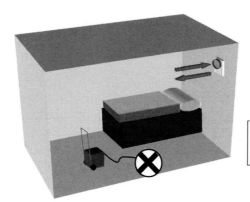

Control (n = 10) – No new bedding
Placebo steam cleaning
Placebo fan

Figure 6.5 Control group intervention measures schematic.

levels and patients' lung function. No attempt was made in any of the groups to measure other changes to internal air quality such as the concentration of CO_2, methane, particulate matter, volatile organic compounds or environmental tobacco smoke, all known to have a relationship with triggering/exacerbating asthmatic symptoms.

Dust collection

As the collection and identification of dust mites has proven to be particularly problematic in other studies, a decision was made to concentrate on measuring changes to the allergen reservoirs. Such a strategy ignores current dust-mite activity and focuses on the level of accumulated allergen. As the HDM produces *c.* 60 times its own body weight in faecal output in its 8–12 week lifespan, a technique for measuring the background level of *Der pI* (*Der pI* is one of the *c.* 15 proteins that have to date been identified in HDM faecal pellets and is considered to be a good marker for measuring total allergen burden) was adopted. *Der pI* can thus be used as a proxy for assessing changes in the available allergen reservoir. Using a conventional vacuum cleaner with a proprietary filter device (supplied by ALK [UK], 8 Bennet Road, Reading, Berkshire, RG2 0QX), dust samples were collected from floors, beds and pillows. A small petri dish with a perforated underside was located in the centre of the filter device which was attached to the vacuum cleaner nozzle. A piece of filter paper placed inside the petri dish filters the airflow before it enters the drum of the vacuum cleaner. The petri dish is then removed and sealed in an airtight bag. An area of approximately $0.5\,m^2$ was selected as a practical size to be vacuumed for 60 seconds – a protocol previously used on other similar studies. Samples were collected from the living room carpet, and all asthmatics' bedroom carpets and beds. All dust samples were collected by the same individual to ensure consistency of method. All dust samples were immediately sealed and chilled (4 °C) after each day of sampling, to prevent any live mite activity occurring in the petri dishes. The vacuum cleaner used for the site work was a simple tub vacuum (Goblin Aquavac – Type 7409P) with no internal filters. The area of carpet to be sampled was randomly chosen in each dwelling. Households were instructed not to vacuum carpets for three days prior to a programmed sampling cycle visit.

Immunological assays

The dust collection system was 'road tested' in several dwellings prior to the initiation of the main project. The dust samples were sent to the Glasgow Western Infirmary's Immunology Department where immunological assays testing for *Der pI* were performed on each sample. Results were returned in the form of micrograms *Der pI* per gram of house dust and absolute amounts were calculated by multiplying the ratio with the individual sample weight. Each assay was performed twice at this stage to verify consistency of the assay kit and procedure. The assay kits were supplied by Advanced Allergy Technologies Ltd (now Alprotect, 3 Millbrook Business Centre, Manchester M23 9YJ).

Der pI is the internationally recognised shortened name of a major allergen type excreted in the faecal pellet of the HDM, *Dermatophagoides pteronyssinus*. It is a protein with cystein proteinase activity and is thought to be a digestive enzyme.

Background and procedures of the assay method

The *Der pI* concentration can be measured by sandwich enzyme immuno-assay. Microplate wells are supplied pre-coated with *Der pI* antibody. Test samples and standards are incubated in the wells and any *Der pI* is bound by the immobilised antibody. The next step involves adding an enzyme-conjugated antibody to *Der pI*. This binds in proportion to the amount of bound *Der pI* and the enzyme activity measured by an enzyme-catalysed chromogenic substrate reaction is a measure of the *Der pI*.

Assay protocol:

1   The dust sample (which had been stored at 4 °C) was transferred from the petri dish into pre-weighed bijoux bottles and the total sample was then weighed.

2   Approximately 200 mg, representative of the total sample, diluted 1/10 dust weight/volume ratio in dust extraction buffer (0.125 M Ammonium hydrogen carbonate buffer + 0.1% sodium azide). The sample was agitated on a rotator for 2 hr at room temperature.

3   The sample was then centrifuged for 15 min at 1 000 times gravity and extracted supernatant stored at −20 °C until required.

4   Each sample extract was diluted 1/10 and 1/100 in sample/standard diluent. *Der pI* standard prepared at 200 ng/ml, 100 ng/ml, 50 ng/ml, 25 ng/ml, 12.5 ng/ml, 6.25 ng/ml, 3.25 ng/ml and 0 ng/ml. Diluent and standard provided by kit.

5   100 µl standard or test sample was added to each well of ELISA plate, covered with plate sealer and incubated for 2 hr at room temperature.

6   The plate was washed three times with wash buffer and 100 µl and biotinylated anti-*Der pI* antibody was added to each well and incubated for a further 2 hr.

7   The plate was washed three times and 100 µl streptavidin conjugated alkaline phosphatese was added and incubated at room temperature for 20 min.

8   The plate was washed three times and 100 µl substrate was added to each well and incubated at room temperature.

9   The optical density of each well was measured with a microplate reader set at 405 nm, when the 200 ng/ml standard reached an absorbance of 1.0.

A standard curve was created and the concentration of mite allergen in the unknown samples was determined by calculating the concentration of *Der pI* corresponding to the mean absorbance from the standard curve, remembering to take into account the dilution factor of the sample i.e. µg/g dust equates to the product of the concentration of the unknown sample (ng/ml) times and ten (ml/g) times the dilution factor of the dust sample.

Environmental monitoring

To identify the longitudinal performance of each house participating in the study, small thermo-graphs (Tinytalk series – Gemini Dataloggers (UK) Ltd) were deployed in all living rooms and thermo-hygrographs in bedrooms to monitor and record the temperature and RH over the duration of the study. In order to achieve an accurate record of any fluctuations of temperature and/or RH within each house, the period between each reading had to be reasonably short. This factor had to be weighed against the storage capacity of the loggers. A 90-min interval between readings was selected for the monitoring programme, as this enabled reasonable coverage of any fluctuations within the homes (e.g. due to showers, cooking, clothes drying etc.) whilst also providing a reasonable timespan of up to 16 weeks before downloading was required. The units were programmed to stop recording when the memory was full to avoid overwriting any previous readings. Two temperature and RH sensors were located externally at two addresses, in positions sheltered from direct exposure from the elements to act as micro-weather stations. The data from these was used to scope any significant variations from local meteorological stations and determine the effect external climate plays on the internal environment.

Energy rating

The identification of every house type involved in the study was undertaken at the initial home visit. Details of materials, construction, orientation, site exposure, dimensions, and heating systems were noted at this stage on a proforma. A software package supplied by NHER (National Home Energy Rating Scheme – National Energy Services Ltd, Rockingham Drive, Linford Wood, Milton Keynes, MK1 6EG) used this information to calculate a SAP and an NHER rating. The NHER method utilises constructional data and details on the type of heating system and its relative efficiency, in addition to the pattern of use and the dwelling's location, to produce a value between 0 and 10. By processing each house type using this software, the values produced for SAP and NHER were compared to national averages to benchmark the relative performance of each house type.

Fuel consumption

The fuel consumption (electricity and/or gas) was recorded for each household. By comparing the fuel consumption and environmental profiles (internal and external) with the NHER and SAP ratings, an estimation can be made to determine whether each house type is over/under performing in terms of energy use throughout the course of each year. Infrared thermography was also used on three dwellings to determine the efficacy of the urea-formaldehyde

Figure 6.6 Thermo-graphic photo showing cold bridging.

cavity fill insulation in a sample of house types. Figure 6.6 shows problems with cold bridges at lintels, sills and reveals, as well as possible uneven fill or slumping below windows. The NHER figure may thus overestimate the dwelling's actual energy efficiency.

Cohort generation

The project costs and budget (funding was assembled from a range of diverse organisations all of which could be considered to be stakeholders in this research area) allowed a target cohort of 80 individuals to be generated. This was initially attempted by letter distribution to individual homes in the target areas. North Lanarkshire Council – as one of the main sponsors – supplied maps for four catchment zones within West Central Scotland which fall under its jurisdiction: Viewpark, Orbiston, Forgewood and North Motherwell; all of which are considered 'social inclusion partnership areas'.

An introductory letter, including a short questionnaire outlining the general entry requirements for participation was developed and delivered to 400 homes in Viewpark along with a Freepost envelope for the return of the completed questionnaires. A reply period of one week was indicated on these letters for all interested parties. The number of leaflets requiring to be delivered for the generation of the 80 participants was then proportionally estimated from the responses received from the initial 400. Less than 10 replies were received. This required a further 4000 letters to be printed and posted. This second wave of letters generated a further 30 volunteers but this was still less than half the target cohort. A new strategy had to be adopted. All primary and secondary schools within the three catchment areas were contacted and asked to distribute leaflets to all pupils in all years. Every school agreed and a further 4 000 letters were delivered. This final campaign, together with the initial leaflet drops generated responses from 68 individuals in 44 dwellings.

Monitoring health status

In order to ascertain whether remedial measure implementation has any beneficial or detrimental effect on the asthmatic subjects, a set of measures were designed to monitor and help quantify the general health of the participants at any time in the study.

Peak flow

Peak flow monitoring is used to track fluctuations in the respiratory health of a patient. Blowing into a plastic tube propels a scaled slider piece and measures lung capacity in litres per minute. Peak-flow meters were supplied to all participants who recorded the best of three morning and evening peak-flow attempts.

Life event diaries

All individuals were asked to record major events in a life event diary. Events to be recorded included: changing of furniture (especially beds and sofas), illness, holidays, prescription changes, emotional distress etc. It was hoped that these diaries could be used to identify any major changes and help to scope confounding variables.

Health questionnaires

To address the practical issues involved in the implementation of this type of dust mite control/avoidance regime, a short questionnaire was devised. This was completed at three-month intervals throughout the study. A hybrid short form was derived using the Euroqol[1] and McMaster[2] questionnaires. As the study involved a large percentage of children it was felt necessary to omit many of the questions developed in the two existing proformas which dealt with purely adult topics.

Prescription profiling

The intention was to audit all patients' medical records to quantify any changes in drug use over a period of four years (two years prior to intervention and two years after remediation). Any changes in drug profile could be used as a proxy for symptom/disease severity. Such profiling could also help with estimating the cost implications of the remedial strategies employed, vis-à-vis capital costs versus the cost of prescription drugs, medical consultations, hospital admissions, work/school days lost and a general quality-of-life index.

If the interventions successfully reduced exposure to indoor dust mite allergens and inhibited re-colonisation rates it was expected that those participants who were dust mite sensitive, were in turn, likely to show a reduction in symptoms and an increase in lung function. As optimum conditions for dust mite proliferation are 25 °C at 80% RH, dwellings with higher temperatures and humidity should show a positive correlation with *Der pI* reservoirs where carpets were of a similar age. If the allergen denaturing and avoidance techniques were efficacious, the quantity of allergen available for inhalation should also reduce. Reducing the dose should lead to positive health impacts. If a person is allergic to a cat, removing the cat and its allergenic by-products from the dwelling would normally result in less symptoms. Dwellings where the absolute humidity has been successfully suppressed (below $7\,g\,kg^{-1}$ of dry air and/or 60% RH) should have slower re-colonisation rates than those dwellings with placebo MHRV units. A comparison of the individual's acute and preventative drug regime against the capital and running costs of the interventions allowed a pay-back period to be estimated.

References

[1] Dorman PJ, Slattery J, Farrell B, Dennis MS and Sandercock PAG. A randomised comparison of the Euroqol and Short Form-36 after stroke. *BMJ*, 1997, Vol. 315, p. 461.
[2] McMaster University. Asthma quality of life questionnaire. Department of Clinical Epidemiology and Biostatistics, McMaster University Medical Centre, 1200 Main Street West, Hamilton, Ontario, Canada, L8N 3Z5.

Chapter 7

Key findings and discussion

Longitudinal measurement of dust mite allergen reservoirs

The initial monitoring cycle confirmed that 55% of living room carpets, 75% of bedroom carpets and 79% of beds contained *Der pI* ratios greater than the WHO sensitisation threshold of $2\,\mu g\,g^{-1}$ of house dust. In addition to this, 18% of living room carpets, 47% of the bedroom carpets and 57% of the beds were found to contain concentrations greater than the upper threshold of $10\,\mu g\,g^{-1}$ of house dust (Figure 7.1).

Comparing the initial levels and levels found six months after the remedial measures were implemented (cycles 1 and 6 – taken during a similar winter period), the percentage above the two WHO thresholds (2 and $10\,\mu g\,g^{-1}$ dust) for all dust samples (carpets and beds) fell from 80 to 21% and 65 to 4% ($n=78$) respectively in active group 2, from 61 to 41% and 34 to 4% in active group 1 ($n=31$) and 65 to 15% in the control group ($n=39$) (Figure 7.2).

Longitudinal measurement of hygro-thermal conditions

Humidity readings taken in bedrooms (every 90 min – the total number of hygrometer readings recorded was in excess of 750 000) demonstrated a reduction in internal absolute water vapour pressure of 12% in active group 2, in comparison with the control group's (measured at the same time of year January–May 1999 vs January–May 2000 see Figure 7.3 for humidity ranges before intervention).

Of the 20 dwellings in active group 2, 17 had Baxi E100 cartridge units fitted in the living rooms and bedrooms. These are balanced units with the air extracted from the room being replaced by an equal volume of external air, pre-warmed by the thin film plastic cross-flow heat exchanger. Three of the dwellings had half-house systems fitted. This regime extracts air from the kitchens and bathrooms and positively pressurises the bedrooms and living room, by delivering fresh pre-warmed air via ducts housed in the roof

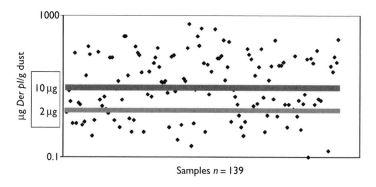

Figure 7.1 Log ratio of initial *Der pI* concentrations (showing WHO sensitisation thresholds).

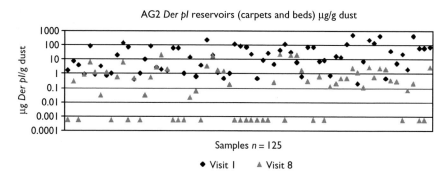

Figure 7.2 Reduction in *Der pI ratios* before and after intervention.

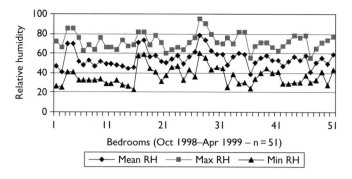

Figure 7.3 Relative humidity ranges before intervention.

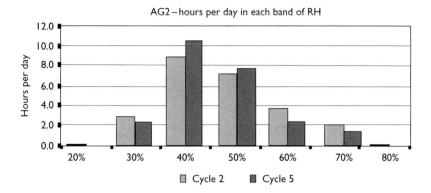

Figure 7.4 Reduction in RH after intervention (active group 2 – all measures).

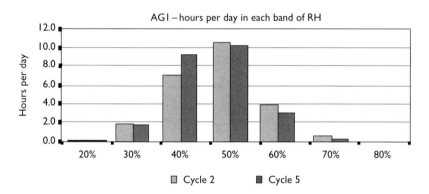

Figure 7.5 RH before and after intervention (active group 1).

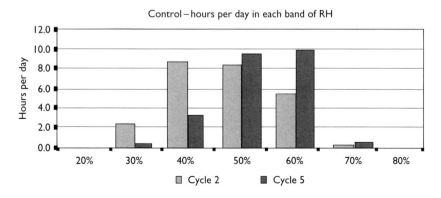

Figure 7.6 RH before and after intervention (control group).

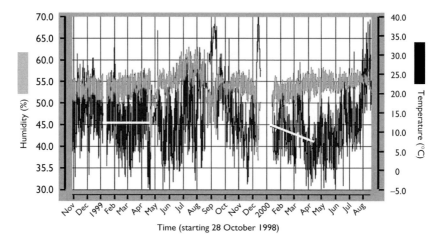

Figure 7.7 Placebo unit RH profile before and after showing trend line.

space. Wet zone extraction, in addition to positive pressurisation, creates a negative pressure in the areas of greatest water vapour production. The thermohyrographic monitoring suggested that such a regime (where the air change rate was also further increased) is more efficacious in reducing internal absolute humidity. Figure 7.7 shows the typical reduction in RH produced by the cartridge unit over a comparable three-month period (January–May 1999 average temperature 5.9°C at 5.0gkg^{-1} dry air, January–May 2000 average temperature 5.4°C at 4.9gkg^{-1} dry air). The unit, reduced the internal RH from 55 to 42% RH. This reduction is likely to be particularly significant in drying out the micro-climates and inhibiting dust mite activity and re-colonisation rates. When the three half-house systems – which extract air from the kitchens and bathrooms and deliver pre-warmed fresh air to the living room and bedroom – were compared with the through-the-wall cartridge (E100 units) a further reduction in internal water vapour pressures is evident (Figures 7.8).

Temperature profiles

Thermographs placed in the living rooms and bedrooms took air temperature readings at 1.5 m above floor level. The total number of temperature readings recorded was in excess of 1.5 million. Figure 7.10 shows that there was no significant seasonal reduction in living room temperatures, whereas Figure 7.11 demonstrates that the increased ventilation rate in the less well heated bedrooms does appear to have reduced the winter temperatures by around 1 °C. Such a reduction would normally result in an increase in

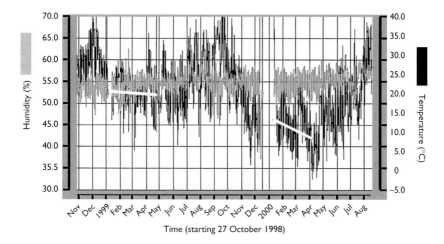

Figure 7.8 MHRV active cartridge unit before and after showing trend line.

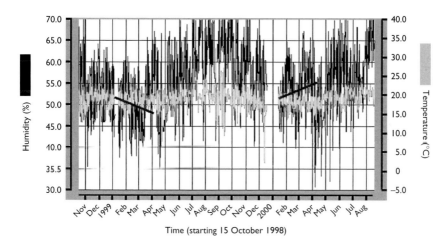

Figure 7.9 Whole-house MHRV loft-mounted unit before and after showing trend line.

internal RH, however, this did not occur as the increase in air change rate has expelled a greater amount of water vapour from the bedroom. Such data can be used to correlate temperature with dust mite activity and measure any drop in internal temperature due to increased ventilation rates. Wet zone extraction in combination with pre-warmed delivery reduced the RH by *c.* 10%. As the internal temperature has remained stable this equates to a significant reduction in the internal water vapour pressures in the range

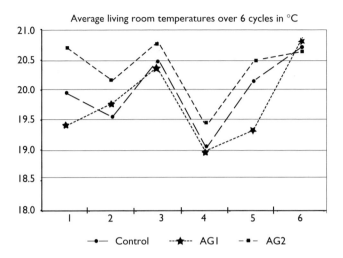

Figure 7.10 Average living room temperature over 6 cycles (all groups).

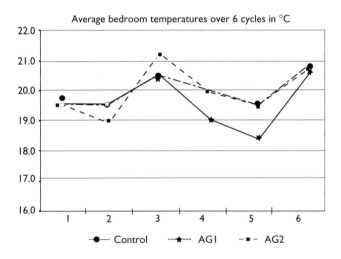

Figure 7.11 Average bedroom temperature over 6 cycles (all groups).

known to inhibit dust mite viability. Figures 7.4–7.6 demonstrate the effect of the MHRV interventions' ability to increase the amount of time RH remains below the 60% critical ambient threshold. Comparing cycles 2 and 5, the units reduced the amount of time the bedroom was above the 70% RH threshold by 52%, and 60% RH by 37%. It has also increased the time by almost 10%, the time duration where the RH was below 50%. E100 cartridge room ventilators were not as effective as the whole-house systems, which had wet zone extraction in addition to positive pressure, but still achieved significant reductions in internal water vapour pressures and improvements in perceived air quality. As expected the placebo fan units had no significant effect on internal water vapour pressures as they were simply re-circulating the internal air. The slight increase in RH (see Figure 7.9) may have been due to the residents reducing the amount of time windows were opened, confirming that these particular fans were convincing placebos.

Changes in health status

Daily peak-flow readings taken before and after the interventions at the same time of year show some lung function improvements in all groups (Figure 7.12). The intervention had the effect of sensitising the entire cohort, and many households in the control groups made lifestyle changes – such as removing carpets – which may have had a beneficial effect driving unexpected improvements in lung function. These figures would also have to be correlated with any changes in drug use before trends could be identified. Some of the increase will be due to child growth.

Health status questionnaires

Face-to-face interviews using a hybrid of the McMaster health questionnaire (McMaster 1989) and Euroqol (Dorman 1997) were completed on a three-month cycle by a contract research assistant blind to the cohort groupings. Tables 7.1 and 7.2 are results from the two key questions introduced for the cycle 6 data collection.

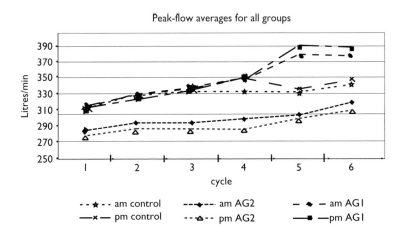

Figure 7.12 Peak-flow averages for all groups.

Q.1 How has the air quality in your home been affected since remedial measure implementation?

Table 7.1 Self-assessed air quality

	Improved	Same	Worse
Control (n=7 houses)	1	5	1
AG1 (n=7 houses)	6	1	0
AG2 (n=20 houses)	18	2	0

Q.2 How has your asthma been affected since remedial measure implementation?

Table 7.2 Self-assessed improvement in asthma status

	Improved	Same	Worse
Control (n=10 people)	1	7	2
AG1 (n=12 people)	5	7	0
AG2 (n=32 people)	26	6	0

GP visits

Access was gained to 31 patients' health records (final cohort $n=54$). The audits were taken after 24 months had elapsed post-intervention and thus a measure could be taken over a four-year period (two-years post- and two pre-intervention) to quantify any significant changes in frequency of GP visits. Where possible, records detailing the total amount and type of pre-scribed asthma drugs were collected ($n=26$). Due to a lack of detail, in combination with the relatively young age of the cohort, it was difficult to generate anything other than illustrative case studies. Table 7.3 represents a 48-month period (averages per person for the 24 months before and after the interventions). Asthma visits were identified by drug prescribing, peak-flow readings or notes mentioning wheeze.

Table 7.3 Audited GP visits before and after intervention

	Total visits		Asthma visits	
	Before	After	Before	After
AG2 (n=17) Ave/person	20.4	13.29	6.41	2.35
AG1 (n=8) Ave/person	9.87	7.62	2.87	1.25
Control (n=6) Ave/person	13.0	7.33	5.16	1.5

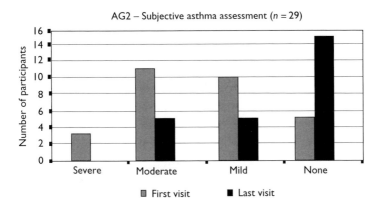

Figure 7.13 Self-assessed asthma status previous week: AG2 (initial and last cycle).

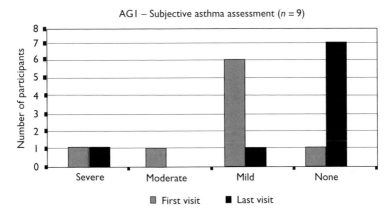

Figure 7.14 Self-assessed asthma status previous week: AG1 (initial and last cycle).

All patients were asked to classify the severity of their asthma in the previous week. Figures 7.13–7.15 demonstrate that the changes did not appear equally across all groups, although by random chance, the active cohort did appear to have more severe symptoms at the outset of the study.

Illustrative case studies

Two patients whose medical records were complete and decipherable for the 48-month period and who had classified their asthma in the initial questionnaire as either severe or moderate, were selected for detailed prescription profiling (Tables 7.4 and 7.5).

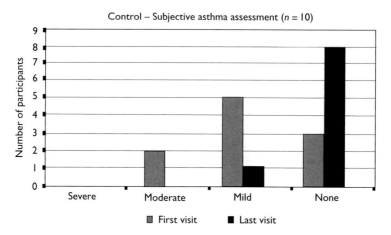

Figure 7.15 Self-assessed asthma status previous week: Control (initial and last cycle).

Table 7.4 Illustrative case study I

Patient 30 (female aged 13)	Before	After	% Reduction
Fluticasone	120 g (£311)	60 g (£156)	50%
Fluticasone (nasal spray)	5 g (£23)	2.5 g (£12)	50%
Salmeterol	21 g (£200)	15 g (£142)	29%
Terbutaline	350 g (£44)	200 g (£25)	43%
Budesonide	140 g (£130)	40 g (£36)	72%
Total drugs in grams	636 g (£708)	317 g (£371)	51%
Total cost saving in drugs based on MIMS[1] September 2001		=£337	
Number of GP visits April 97–March 99 = 8 at £20		=£160	
Number of GP visits April 99–March 2001 = 5 at £20		=£100	
Total saving on medical costs = £337 + £60		=£397	
Total cost of intervention		=£492	
(Pay-back period < 28 months)			

Electricity and gas profiles

Electricity and gas meter readings were recorded during each monitoring cycle. The data that has been graphed is based on two similar winter periods (October 1998–May 1999: 203 days and August 1999–April 2000: 242 days). The consumption has been converted into kW hours per winter-day period. The graphs are thus only a simple comparative measure and undoubtedly overestimate the total annual consumption (Figure 7.16). Problems did

Table 7.5 Prescription profiling AG2 case study

Patient 11 (female aged 50)	Before	After	% Reduction
Salbutamol	700 g (£67)	520 g (£500)	31%
Beclamethasone	875 g (£621)	650 g (£462)	31%
Total drugs in grams	1575 g (£688)	1170 g (£512)	31%
Total cost saving in drugs based on MIMS[1] September 2001 = £176			
Number of GP visits April 97–March 99 = 17 at £20		= £340	
Number of GP visits April 99–March 2001 = 4 at £20		= £80	
Total saving in medical costs = £176 + £260		= £436	
Total cost of intervention		= £492	
(Pay-back period < 27 months)			

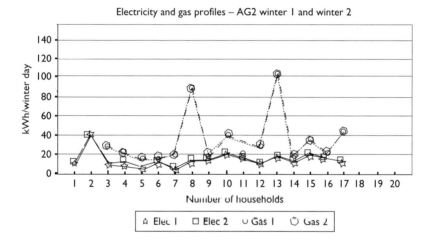

Figure 7.16 Mean fuel usage in AG2 kWh/day over 2 similar winter periods.

arise where households had changed their supplier and new meters were fitted, which resulted in some data loss.

Discussion

Initial Der pl *levels*

In general, there has been little research done into dust mite colonisation of the housing stock. This is in part due to the problems of identifying and trapping live mites. As mites produce 60 times their own body weight in faecal pellets (which contain at least 16 identified allergenic proteins) and, as these proteins do not biodegrade, they can be used as an accurate longitudinal marker to assess historical dust mite colonisation and activity. The level of *Der pI* identified in the existing dust reservoirs was significant, with 55% of the living room carpet dust samples, 75% of the bedroom carpet dust samples and 78% of the bed surfaces harbouring levels above the WHO threshold for allergic sensitisation in atopic individuals. Indeed, 50% of the bedroom carpets and 56% of the bed surfaces contained more than 10 μg/g, the upper threshold known to trigger allergic symptoms. Since the amount of dust mite allergen seems to have an associated clinical risk – that is, a dose–response relationship – it is noteworthy that some exceptionally high concentrations were found, with 18 samples above 100 μg *Der pI*/g of dust, 9 of which were above 200 μg *Der pI*/g of dust and 3 of which were above 400 μg *Der pI*/g of dust. There was thus a significant number of households where the level of HDM allergen was high enough (60 μg/g) to sensitise non-atopic individuals.

This study has identified that eight out of ten cohort subjects were being exposed to HDM allergen reservoirs above the WHO sensitisation threshold of $2 \mu g/g^{-1}$ of house dust (for atopic individuals) in one or more locations in their dwelling. Although the cohort consisted of volunteers, the house types studied are typical of the Scottish public-sector stock as a whole. Although caution must be exercised when extrapolating from such a relatively small cohort, however, as these dwellings are not atypical of the public-sector stock in Scotland – or for that matter the UK as a whole – the results clearly give rise to concerns and justify further and more comprehensive investigations.

Efficacy of active measures

Comparing the first and last sampling cycle in the active group – which had the additional MHRV intervention – absolute levels of *Der pI* dropped to 7% (all carpets) and 2% (beds), of their original ratio (October 1998–September 2000). This compared with 66% (all carpets) and 5% (beds) for active group 1. From this we can conclude that the barrier bedding was effective in inhibiting *Der pI* escaping from the mattress. No useful comparison could be made with the other control group due to low sample numbers achieved in cycle 8. The graphs of dust sample median weights suggest that dust is not equally distributed across a room and thus care should be taken when results are expressed as ratios. The asymmetry of the distribution may actually be larger than the (before and after) differential being measured. No pattern emerged in terms of median dust weights, which suggests that the dust may not be evenly distributed throughout a room. The precise location of the cyclical dust sampling, can drive major variations in both dust weight and *Der pI* concentrations. Figures 7.17 and 7.18 were an attempt to quantify changes in the total weight of dust collected with the

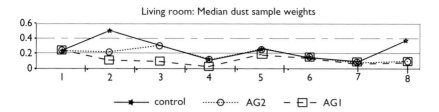

Figure 7.17 Median weight of dust samples across eight cycles (living room).

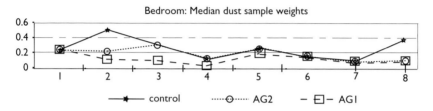

Figure 7.18 Median weight of dust samples across eight cycles (bedroom).

absolute amount of *Der pI* in the sample. Even when the median value is used – in an attempt to reduce the asymmetry of the spread – no clear picture emerges which differentiates the cases from the controls. It is clear that the *Der pI* concentrations drop across all groups even though the medians of dust weight do not show the same proportional changes. This is clearly problematic. Why should the *Der pI* concentrations reduce in the controls even though the dwellings do not appear to be any cleaner in terms of median dust weights? There was some anecdotal evidence to support the hypothesis that the control cohort became sensitised and this in turn generated lifestyle changes such as more intensive cleaning, the use of new vacuums incorporating HEPA filters and the fitting of timber laminate flooring. It appears that a new sampling protocol requires to be developed that can accurately assess the total weight of allergen available for re-suspension and inhalation. Sampling a relatively small area ($0.5\,m^2$) can produce an unacceptable level of background noise.

The MHRV units, by providing *c.* an additional $1\,ach^{-1}$, reduced the absolute humidity in the bedrooms of the active group by just over 12% ($0.42\,g\,kg^{-1}$ of dry air – cycles 2 and 5) when compared with the control groups. Figures 7.7–7.9 show individual house profiles that best demonstrate these trends. This may have inhibited re-colonisation and accounts – at least in part – for the difference in the carpet sample results. Thermographic profiles and data from the whole-house MHRV units, confirmed that wet zone extraction, in addition to preheat positive input to bedrooms and living rooms, further reduced the absolute humidity.

Temperature and fuel expenditure profiles

The data sets confirmed that there had been no significant drop in the average living room temperatures in the active group, despite the increase in background ventilation, however, the average bedroom temperatures dropped by almost 1 °C to just over 18 °C (see Figure 7.11). Clearly there is a greater heat input into the living room over an extended period and thus more heat can be reclaimed. As the units have a low-power consumption (15 Watts), the annual running cost is c. £7.50. A before-and-after audit of fuel expenditure in active group 2 shows no significant variation (kwh/winter day – see Figure 7.16).

Peak-flow and health questionnaires

It is clear from Figure 7.12 that all groups appear to have benefited in terms of the increase in daily peak-flow readings. Part of this increase will simply be due to child growth. What becomes obvious from studying the immuno-logical assay data sets, is that both control groups also appear to have reduced their exposure levels. It is difficult in such a prolonged and cyclic intervention not to sensitise the control group to what is considered to be the main source of allergen. Repeated contact may have brought about a process of education and greater awareness, which in turn, may have driven subtle changes in behaviour. Several children in the control group were also suffering from colds during the first visit, which may have produced unusually low initial readings.

The health questionnaires identify some significant variations between the groups, with 18 out of 20 households in AG2 reporting improvements in air quality and 26 out of 32 individuals in the same group reporting improved lung function (see Tables 7.1 and 7.2). The two control groups did not match these improvements. From these figures we can conclude that there was a highly significant association between the extent of the intervention and perceived improvement in both air quality (Fisher's exact test $p = 0.001$) and asthma (Fisher's exact test $p < 0.0001$). Testing at 5% level of significance, Q.1 demonstrates that the three groups do not have the same probabilities for improvement. The active groups cannot be separated (at the 5% level) but have higher probability than the control group. The estimated probabilities are: 14 and 89%. Again, for Q.2 the three groups are not the same. The control cannot be separated from AG1, with an estimated common probability of improvement of 27%, whereas AG2 achieved 81%. Although the participants were unaware that some fans were placebo units – which simply re-circulated the internal air – the health questionnaires demonstrated significant improvement in air quality and asthma.

Subjective individual assessment as to the severity of their disease in the previous week, again confirmed significant improvements with no individual in the active group (Figure 7.13) now classifying themselves as having 'severe' asthma and 15 of the cohort reporting that their asthma was 'non-existent'.

GP visits and drug profiling

Mandates from 31 patients were received and a full audit of their medical records over a four-year period (two years before and two years after inter-vention) was undertaken. The number of visits to their GP for all ailments was recorded, with asthma-specific visits recorded separately. Although the medical records were adequate to construct simple attendance records, the prescription histories were invariably incomplete or missing. It was thus only possible to produce illustrative individual case studies, both of which confirm the cost-effectiveness of the intervention for these particular indi-viduals (see Tables 7.4 and 7.5).

The GP visit figures show across-the-board benefits, with all groups reducing attendance for both all ailment and asthma-specific visits. Despite the random distribution of individuals into the various groupings, by randomised chance, AG2 appeared – from the outset – to be significantly unhealthier than the control groups, with patients attending the GP almost twice as many times (10.2 visits per year compared with 5.7). Although the figures appear to demonstrate that the control group had similar health benefits, this is due in large part to attrition and the numbers fell below the threshold of significance.

Confounding variables

The methods used were unable to provide information on the following areas.

Air infiltration rates

The background air infiltration rates of the dwellings in the study were not tested. The fan units provide – in ideal conditions where the bedroom/living room door is kept closed – an additional 24 air changes per day. This has to be viewed in the context of the natural ventilation and air infiltration characteristics of the dwelling. As previously discussed in Chapter 5, air leakage rates can vary significantly from dwelling to dwelling. Any changes to the ventilation regime can only be properly evaluated when taken as a proportional component of the existing background air infiltration rate.

This, however, is a relatively expensive and time-consuming exercise. Any future study must attempt to scope this variable if the efficacy of the fans is to be more accurately measured or estimated.

Possible filtering effect of placebo fans

Of the 44 original dwellings in the study, 22 had placebo units fitted. These were designed to mimic the active fans by re-circulating the internal air. The filters incorporated in the units may have had a significant air scouring effect. The fans may not have been ideal placebo units as they may have had some influence on airborne allergen levels. Any future study must reduce this background noise by eliminating or fully incorporating any filtration effects. Such a filtration effect may also change through time as the filter becomes caked with dust and allergen. As convincing placebos, the individuals in the control groups may also have relied on the fans to provide background ventilation and reduced the frequency of window opening. This may account in part for the increase in indoor RH during the winter months in the control groups (see Figure 7.9). When compared with the active fans, the difference may well be significant for dust mite proliferation rates or even colony viability.

Childhood asthma and HDM sensitisation

The cohort was assembled by delivering invitations via all primary and secondary schools in the area of North Lanarkshire. The age profile was thus heavily skewed with 65% of the sample being under 16 at the outset of the study. As child asthma may be more capricious and less chronic than adult cases, a study that concentrates on patients with a higher degree of documented clinical history may allow more accurate measurement of any changes in lung function. As none of the volunteers had undergone skin prick tests to assess their sensitisation to the HDM, it is not known how many would directly benefit from lowering exposure to the allergenic proteins. Morrison-Smith et al.[2] suggest that c. 80% of allergic asthmatics react to extracts of mite allergens and thus it can be assumed that a similar percentage of the active group will have measurable improvements in lung function and health status. Conversely, 20% of the active group could not be expected to have symptomatic improvements if the only change was in the airborne HDM protein concentrations.

Distribution of Der pI

The initial three dust sample cycles provided evidence that *Der pI* is not evenly distributed throughout a room. The protocol dictated that a different area of carpet be sampled at each cycle, however, the results clearly indicate that *Der pI* concentrations or absolute weights, can vary significantly across a carpet. This asymmetry demands new sampling techniques be developed to ensure that a more accurate assessment of the average total allergen burden contained in the dust reservoirs is extracted at each monitoring cycle. The background variation across a surface may actually be greater than the differential being measured, before and after remediation.

Cohort lifestyle changes

Any interventionist studies run the risk of educating the cohort into lifestyle changes. Constant monitoring and dust sampling clearly affected many of the participants, some of whom made alterations to indoor floor coverings (removing carpets and introducing timber laminate flooring) and/or cleaning regimes (vacuums with HEPA filters, purchasing steam cleaners). When the participants were in one of the control groups, this could well have influenced their exposure rates and hence the data (lung function and GP visits) appears to demonstrate that all participants have benefited. Despite encouraging the cohort to complete life event diaries, noting any significant changes in lifestyle or domestic furnishings (new bedding/suites/pets etc.), only one such diary was completed and returned out of 68 dispensed.

Summary

The initial monitoring cycle confirmed that 55% of living room carpets, 75% of bedroom carpets and 79% of beds contained *Der pI* ratios greater than the WHO sensitisation threshold of $2 \, \mu g \, g^{-1}$ of house dust. In addition to this 18% of living room carpets, 47% of the bedroom carpets and 57% of the beds were found to contain concentrations greater than the upper threshold of $10 \, \mu g \, g^{-1}$ of house dust.

The average bedroom humidity measured over the first three months at the outset was remarkably consistent across all dwellings at *c.* 55% RH. This represented a mixing ratio between 7.5 and $7.8 \, g \, kg^{-1}$ of dry air. The internal humidity was clearly being influenced by external conditions with both the RH and AH dropping during the three coldest winter months by about 3% and *c.* 0.5 g of water vapour. There was no significant relationship between RH measured over this relatively short pre-intervention period and the level of *Der pI* in the bedroom dust reservoirs.

Comparing the initial levels and levels found six months after the remedial measures were implemented (cycles 1 and 6 – taken during a similar winter period), the percentage above the two WHO thresholds (2 and $10 \, \mu g \, g^{-1}$ dust) for all dust samples (carpets and beds) fell from 80 to 21% and 65 to 4% ($n = 78$) respectively in active group 2, from 61 to 41% and 34 to 4%

in active group 1 ($n=31$) and 65 to 15% in the control group ($n=39$). The remedial measures can thus be considered reasonably effective in reducing allergen burdens in the dust reservoirs. Effective allergen denaturing/avoidance techniques such as steam cleaning and mattress encapsulation can significantly reduce the reservoir of allergenic protein available for inhalation.

The MHRV cartridge units reduced internal AH when compared with the two control groups by $c.$ 12%. MHRV units can reduce internal water vapour pressures in this region – particular region, particularly during the winter months – to a level below $7\,g\,kg^{-1}$ of dry air and/or 50% RH, which is likely to inhibit dust mite activity and viability. More significantly, the units were able to reduce the amplitude of the diurnal variation, by reducing the amount of time the bedroom was above 70% RH by 52%, and 60% RH by 37%. The units also increased the time where the RH was below 50% by almost 10%. No significant changes were noted in the control groups.

Although all groups' lung function appeared to improve by a margin of between 20 and 40 l/min, some of this increase will be accounted for by child growth. There was also no significant difference in peak-flow data between the cases and the controls, with all groups apparently benefiting from the trial. The health questionnaire data, confirmed that there was a highly significant association between the extent of the intervention and perceived improvement in both air quality (Fisher's exact test $p=0.001$) and asthma (Fisher's exact test $p < 0.0001$). Testing at 5% level of significance, Q.1 demonstrates that the three groups do not have the same probabilities for improvement. The active groups cannot be separated (at the 5% level) but have higher probability than the control group. The estimated probabilities are: 14 & 89%. Again, for Q.2 the three groups are not the same. The control cannot be separated from AG1, with an estimated common probability of improvement of 27%, whereas with AG2 it is 81%.

Although the prescription records were not complete, there was a significant drop in GP visits. The two illustrative case studies calculated a pay-back period of $c.$ 27 months. These two patients were, however, on high steriod and Beta$_2$-agonist doses. If the payback is based on cost benefit and an account is made for lost workdays etc., it is highly likely, that, in addition to the self-reported clinical improvements, significant cost savings from a reduction in drug use and primary/acute care services can accrue, rendering such an approach cost-effective in the medium term, both as a treatment and a preventative strategy.

The study has demonstrated that remediation measures which denature/ encapsulate dust mite allergens in combination with increased air change rates, can result in a reduction in allergen exposure, an improvement in perceived air quality and lung function and a suppression of internal water vapour pressures which should inhibit dust mite re-colonisation and activity.

References

[1] MIMS On-line (drug costs) http://www.mims.hcn.net.au/.

[2] Morrison-Smith J, Disney ME, Williams JP and Goels ZA. Clinical significance of skin reactions to mite extracts in children with asthma. *BMJ*, 1969, Vol. II, pp. 723–726.

Chapter 8

Scoping the confounding variables

There are a number of factors and confounding variables that can provide a degree of background noise. Any future research studies should be designed to reduce the influence of the following confounders:

(1) What is the minimum cohort number required to produce a statistically significant outcome?
(2) What is the natural variation in *Der pI* distribution across a carpet; what are the factors driving this variation and can the dust sampling technique be modified to take account of any asymmetric *Der pI* distribution?
(3) What are the factors driving variations in the ratio of airborne allergen to allergen contained in the dust reservoirs?
(4) What is the likely influence of HEPA filters and more powerful vacuum technology on dust and allergen reservoirs?
(5) What additional clinical tests can be introduced to identify dust mite sensitivity, quantify antibody generation in the bloodstream and measure changes in lung function and medication?
(6) How best can economic outcomes be measured?
(7) What is the role of air infiltration/leakage rates in enhancing or inhibiting the MHRV intervention?
(8) What is the significance of other indoor air pollutants such as gases, microbes, moulds, particulates, endotoxins and toxic chemicals on the asthmatic state and which ones can be scoped or accurately measured to provide secondary outcomes without compromising any studies' primary objectives?

Questions 1–7 will be dealt with in this chapter while the evidence base covering the main aspects of Question 8 will be reviewed in Chapter 9.

Statistical power calculations

If the intended power of a study is to be 80% (at the 5% level), to detect a mean treatment difference of $20 \, \text{l min}^{-1}$ in peak expiratory flow (PEF) as a primary endpoint, with a standard deviation in PEF of $40 \, \text{l min}^{-1}$, a sample size of 64 cases and 64 controls – to be matched where possible for age, sex and smoking history – will be required. If there is unusable data from 10% of subjects i.e. only 115 patients complete the study, then to maintain the power at 80%, the mean treatment difference would be $22 \, \text{l min}^{-1}$ in PEF. As some decay in the sample size is predictable – based on the initial research programme where a proportion of the residents moved during the two-year study period – a target cohort of 140 will be required to allow for some attrition. The primary analysis should be a comparison between groups, of the change over baseline in morning lung function. Secondary endpoints such as symptom scores, exacerbation rates, spirometry, quality of life with respect to asthma, *Der pI* levels in the homes, sensitisation levels, economic evaluations and humidity readings could also be generated. The main analysis could be carried out using normal linear models that adjust for baseline severity. Other outcomes may use Wilcoxon tests of differences from baseline or logistic regression with adjustment for baseline. All analyses should be carried out on an intention to treat basis.

Der pI distribution and sampling method

A recent study by Macher *et al.*[1] in the United States which measured mite allergen levels in 93 commercial office buildings, found that only five had levels above the WHO sensitisation thresholds. Although mite allergens are not exclusive to the domestic environment, it is highly likely that the greatest exposure takes place in the home. There are a number of studies that show dust mite allergens to be asymmetrically distributed in domestic dust reservoirs.

Four recent studies[2–5] which examined the distribution of dust mite allergens within individual rooms, found significant variations among the separate samples collected. In one study a 24-fold difference in distribution was found in bedroom carpets. Such variation throws doubt upon the reliability of any longitudinal sampling method, as the background variation in distribution could actually be larger than the differential being measured before and after steam cleaning or high efficiency vacuuming. It has now become increasingly evident that the habitats of living mites, the site of the initial faecal pellet deposition and the re-distribution of allergens – which have been temporarily re-suspended – are not necessarily coincident.

What is driving this asymmetry in allergen concentrations remains a mystery. It can be hypothesised that proximity to a door may be significant as greater traffic levels can produce more food source, however, it could equally be claimed that concentrations will reduce as these areas are likely to be more heavily polluted with trafficked sand and debris. Proximity to a bed might be associated with more settled aeroallergens emanating from the pillows and mattress, while draughts or periodic air currents driven by convection, may deposit airborne particles in specific niches. As the mites themselves are known to be photophobic, it is possible that they will avoid areas where sunlight regularly falls. There may be no relationship between where the dust mites congregate, where the faecal pellets are re-distributed and where specific mite proteins are found, especially if the pellets break down into smaller and lighter particle sizes.

An additional problem is also generated by using concentrations, rather than absolute quantities of *Der pI*. The majority of studies have used such a measurement protocol, however, knowledge of the absolute amount and relative distribution will be more relevant when attempting denaturing or avoidance interventions, as sites that contribute quantitatively to the total load should be targeted. A room with relatively low dust pollution can appear to have a high ratio of HDM allergen, however, when the sample weight is multiplied by the ratio, the absolute amount per square metre can be negligible compared with a heavier dust sample reported with a nominally lower ratio.

In a recent Danish study by Sidenius *et al.*[6] investigating eight dwellings inhabited by dust mite sensitive asthmatics, the mattresses were identified as significant allergen 'factories' with median concentrations, derived from both *Dermataphagoides pteronyssinus* and *Dermataphagoides farinae*, of $86\,\mu g\,g^{-1}$ of dust (range 30–288) which equated to 1888 live mites per gram of dust (range 12–1910). This compared with $32.1\,\mu g\,g^{-1}$ for the bedroom carpet, $3.2\,\mu g\,g^{-1}$ for the living room carpet and $4\,\mu g\,g^{-1}$ for upholstered furniture. Although the paper highlighted the greater concentration in the mattress, a closer look at the absolute amount of allergen available per square metre of surface area, confirmed that the carpet was a larger and more significant reservoir. The mattress results gave a median value of $31.4\,\mu g\,m^{-2}$ (total median weight of dust 0.83 g) for the mattress surface area. This compared with $19.7\,\mu g\,m^{-2}$ (total median weight of dust 2.45 g) for the bedroom carpets. When the total surface area is multiplied by this ratio, the mattress makes an estimated contribution of $89.7\,\mu g$ against $177\,\mu g$ for the carpet.

Similarly when mites were counted, the median value was similar for both the mattress and carpet at 188 and 210, however, when multiplied by the total area, the study estimated 207 mites on the mattress surface with 1158 in the carpet. Although the results are of great interest, the study recognises that the allergen reservoir itself may not be as significant as an assessment of the likelihood of reservoir disturbance and the release of HDM allergens into the respirable atmosphere. The study concluded that dust from walls, uncarpeted floors, bookshelves and curtains do not appear to contribute any significant amounts to the total HDM allergen load. Denaturing/avoidance interventions should thus concentrate on carpets, mattresses, bedding and soft furnishings. This study generated two further interesting conclusions and hypothesis. Firstly, more than half the homes were dominated by either *D. pteronyssinus* or *D. farinae*, suggesting that one species may be able to dominate the other depending on local conditions. Secondly, the proposed equivalence ratio, previously postulated by Platts-Mills *et al.*[7] of 10 μg *Der I*/g of house dust (*Der I* is the sum of *Der pI* and *Der fI*) which equates to 5000 mites per gram of dust has been underestimated by an order of magnitude. The study found tenfold more allergen per mite, when results from acarological counting and allergen measurement were pooled. There does not appear to be a reliable universal conversion factor and it is highly likely that the age of the dust mite population plays the crucial role in determining the amount of accumulated allergen.

The role of vacuuming and the type of vacuum used may also be significant in pellet and protein distribution. These allergenic particles are relatively large and will rapidly fall in undisturbed air.[7,8] In time, the faecal pellets may be subject to further decay (depending on room activity) and these desiccated fragments – which will be lighter than the pellet as a whole – could become more easily airborne and remain in suspension for a greater length of time. Vacuums without HEPA filters are likely to allow PM under 10 microns to pass through and be distributed by the outgoing air stream. This could lead to a more even distribution across any room. The new generation of vacuum cleaners make impressive claims as to their efficacy in reducing such allergens. A study by White and Dingle[9] which undertook a vacuum intensity of 4 min m^{-2} followed by regular moderate intensity (1 min m^{-2}) reduced – over a period of 14 weeks – coarse PM by 65% and fine PM by 40%. With a reduction in the dust reservoir entrained in the carpets, airborne PM mass concentrations also fell by 73% PM_{10} and 60% $PM_{2.5}$. Mean mass recovery between the first and second clean (14 days) dropped from 11.7 g m^{-2} of carpet, to 5.9 g m^{-2}, with a further drop to 2.9 g m^{-2} at the third clean. At the end of the study this level of intensive vacuuming had reduced total airborne PM close to outdoor ambient concentrations at 21 μg m^{-3} (PM_{10}) and 14 μg m^{-3} ($PM_{2.5}$).

Kemp, Dingle and Neumeister[10] also observed an 80% reduction in an office building in Perth, Western Australia after conventional carpet cleaning was replaced by high efficiency vacuuming. From these and other similar studies, it is clear that aggressive high performance vacuuming can play a part in reducing and maintaining PM of a size that may contain a significant proportion of the HDM allergens available and likely to be suspended if disturbed. Although the market penetration of high performance vacuum cleaners is increasing, it must be recognised that this level of vacuum intensity is not likely to be replicated in the domestic environment. This was a demanding coverage rate and would require an average living room of *c.* 20 m^2 to be vacuumed for 80 min; an unlikely scenario.

Mitakakis, Mahmic and Tovey[11] suggest a protocol that relies on testing four non-adjacent 0.25 m^2 areas that are equidistant from the door. They claim that this will better scope the distribution and an average figure can then be determined from the four separate samples. Such an approach increases the number and cost of each sampling cycle by a factor of four. The challenge is thus to develop a robust protocol which can more accurately measure the changes in the total allergen burden in any given room, without incurring unmanageable cost and time increases.

Developing a new sampling protocol

Any attempt to measure the total *Der pI* as a marker for HDM activity must face the problem of assumed unequal distribution. It is thus obvious that the greater the number of samples, the closer the result will be to the mean and average distributed concentration, however, assays are both time-consuming and expensive, and taking a large number of samples in each room can be ruled out on cost grounds alone.

A problem with filter interface clogging has also been postulated, with vacuum efficiency being progressively compromised. The task is thus to find a filter mechanism fine enough to trap the dust and *Der pI*, but with a weave characteristic that allows good airflow. Alprotect Ltd (Alprotect, 3 Millbrook Business Centre, Manchester M23 9YJ) have developed such a fabric for barrier bedding and are now using it as proprietary 'dust socks' which can easily fit within the neck of a standard vacuum hose nozzle. A technique that sampled several areas using this product could provide a sample that was closer to the total average *Der pI* burden in the dust reservoir.

Comparing dust sampling techniques

The following protocol was tested in eight dwellings (eight bedrooms and four living rooms); four selected from the phase I cohort known to contain relatively high *Der pI* concentrations and four randomly identified.

(a) Four non-adjacent areas of $0.25\,m^2$ were taped off and vacuumed for 30 seconds (Dyson upright) as recommended by Mitakakis, Mahmic and Tovey[12] using the specialised 'dust sock' inserted into the open hose tube.

(b) Ten separate strips of carpet $1000 \times 100\,mm$ wide outwith the taped areas were vacuumed using the upright detachable hose. A single pass was used with the head in contact with the carpet for 12 seconds. In total an area of $2\,m^2$ was tested in 12 rooms of between 14 and $20\,m^2$. Over 90% of the carpet surface area thus remained untouched.

(c) Finally, the entire exposed carpet area was vacuumed for $2\,min\,m^{-2}$ and the contents of the collector vessel decanted into a sealable plastic bag. Four samples of fine dust approximately $1\,g$ in weight were extracted in the laboratory from the total volume and assayed separately.

(d) The area of carpet under the bed was vacuumed for $5\,min$ to identify if this was a preferred micro-climate of this photophobic species.

(e) Total mattress surfaces were also vacuumed for $2\,min$.

Test results

Using the three different vacuuming techniques outlined above and two dust extraction/assay methods (dust sock and four sub-samples form the total dust mass), a total of 124 samples were assayed representing twelve rooms in eight separate dwellings (as Mitakakis, Mahmic and Tovey[11] – eight bedrooms and four living rooms). Four of the rooms were found to have relatively low concentrations with no sample approaching the $2\,\mu g\,g^{-1}$ sensitisation threshold. The distribution in the various samples taken in the eight remaining rooms ranged from: (i) 0.8–$55.3\,\mu g\,g^{-1}$, (ii) 0.9–$30.4\,\mu g\,g^{-1}$, (iii) 2.6–$96.6\,\mu g\,g^{-1}$, (iv) 0.5–$31.4\,\mu g\,g^{-1}$, (v) 0–$8.1\,\mu g\,g^{-1}$, (vi) 2.7–$17.1\,\mu g\,g^{-1}$, (vii) 0–$5.6\,\mu g\,g^{-1}$ and (viii) 0–$3.6\,\mu g\,g^{-1}$. A variation of almost 70-fold in room (1) and 63-fold in room (4). In three rooms the small-area sampling protocol did not register any *Der pI*, however, the sub-samples taken from the large dust mass collected a significant quantity.

Two of the eight mattress dust samples had no detectable ratio. The remaining six were measured at: 0.2, 0.9, 1.9, 2.2, 4.4 and $5.6\,\mu g\,g^{-1}$, however, when converted to absolute weight per square metre, these figures changed to 0.44, 0.8, 0.7, 1.1, 8.4 and $1.51\,\mu g$ per area vacuumed. When compared with the carpet, the bed surface had relatively little *Der pI* contamination. More interestingly, the area directly beneath the bed had also relatively low contamination compared with the exposed carpet. It has been previously hypothesised that the mattress – due to its ability to absorb moisture from sleeping torsos – was a significant factory with faecal pellets being periodically exuded. When combined with the photophobic nature of the species, the area of carpet underneath the bed appears to be an ideal micro-climate. These examples contradict this hypothesis, with the average allergen burden both as a ratio and an absolute amount, being much higher in the surrounding exposed carpet. Four sofas were also tested with three having no detectable ratios and one measuring $0.8\,\mu g\,g^{-1}$.

When the total dust burden extracted from the exposed carpet area was used as a reservoir for four sub-sets, both the ratios and the absolute amount of *Der pI* still varied by up to a factor of five. It appears that even the turbulent action of the vacuum does not produce an even distribution of allergen within a given dust sample. The sub-sample results did fall within a closer range, and thus the method can be considered to be an improvement on the alternative small-area sampling protocol.

Although such an approach can reduce the variation, taking the entire contents of the vacuum drum and adopting a 'bucket chemistry' approach (immersing the entire contents of the drum in saline buffer to dilute the total volume of soluble protein) will produce one mean result, which can be expressed both as a ratio and as an absolute weight per square metre. Vacuuming the entire exposed carpeted area of a room for $1\,min\,m^{-2}$ will extract a significant quantity of the dust mass and allow the results to be expressed both as a ratio ($\mu g\, g^{-1}$ of dust) and as an absolute weight per area ($\mu g\, m^{-2}$) thus reducing the margin or error generated by both asymmetric allergen distribution and variable dust masses.

As the White and Dingle[9] paper supports the view that this level of intensive vacuuming can extract 65% of coarse and 40% of fine PM, it can be presumed that the *Der pI* is extracted at a similar rate across the entire carpet area. Such an approach will more accurately assess the total allergen burden, but will compromise longitudinal studies that do not involve interventions, as the sampling method is itself a reasonably effective eradication technique which will reduce the total allergen reservoir.

Airborne allergen particulates

The relationship between allergen reservoirs and airborne protein concentrations is not well researched. A freshly excreted dust mite faecal pellet is normally in the range of 10–20 microns with a weight between four and ten micrograms. This is a relatively heavy particulate, and in still air will not remain in suspension for much over 30 min.[11] The pellet, however, is dry and friable and may be broken down by vacuuming and foot traffic into much smaller pieces. Little research has been carried out into whether the faecal matter biodegrades, leaving the more robust proteins free in the dust reservoir. As the proteins are below the one micron threshold, there is some uncertainty as to whether such small particles will break the surface tension in the lung, and although these particle sizes can be respired deep into the lungs, they may not be absorbed. The allergenic protein may thus need a matrix of heavier faecal detritus as a vector, if they are to lodge and be absorbed in the lung.

The weight of the allergen is important in determining airborne concentrations. The concentrations in air will be a function of the total volumetric allergen available, the speed and direction of internal air currents – possibly driven by domestic activity (e.g. children playing) or convection – and floor surface characteristics (i.e. type of carpet/linoleum/floorboards). Some researchers have claimed that carpets, although providing suitable microclimates for mite colonisation and activity, can actually play a role in inhibiting airborne allergen distribution by entrapping the faecal pellets in the carpet pile. As the only route into the lung is by airborne particulates it is important to develop an understanding of the factors that affect dust reservoirs and airborne concentrations. It may not be the absolute quantity of antigen in the reservoir that is of prime importance, rather the availability of allergenic particulates in the respirable environment. Factors such as settling and re-suspension rates, multiple sources, turbulence and affinities for interior surfaces make it a challenge to accurately quantify exposure. Braun *et al.*[12] using computational fluid dynamics (CFD) modelling, claimed that carpets can act like a filter. Airborne particles settle into the carpet due to the pull of gravity and are not easily removed by air currents within the room or even by the action of walking. Hard surfaces, on the other hand, offer no additional resistance to air motion and consequently particles may be more easily suspended into the breathing zone.

In a study by Brown et al.[13] in Australia, carpets were found to contain mite populations of 25 000 mites/m^2 in bedrooms and 50 000 mites/m^2 in living rooms. It is of course significant that living rooms had a greater level of contamination than bedrooms, where mattresses have historically been considered to be the main allergen 'factory'. Furthermore a key study by Gunnarsen, Sidenius and Hallas[14] in Denmark, found significantly more mites on the floor than in the beds, with on average, over 15 times more mites found on carpeted than on lacquered wood floors (3 mites/m^2 against 46 mites/m^2). As the mites themselves are known to be photophobic, it is highly likely that they will avoid areas of sunlight. As hard floors provide no shade it is not surprising that live mites migrate away from such an environment. When mite concentrations on mattresses were correlated with four different floor types, no significant difference could be detected (87 apartments). This again supports the hypothesis that dust mite activity in carpets is essentially independent of dust mite activity in bedding and contradicts the assertion that mattresses and soft furnishings are a significantly greater source of allergen than carpets.

Carpet mite activity appears to be variable as it may be dependent on the type of carpet. Carpets should not be considered as a homogenous group of floor coverings as there are significant differences in weave type, face weight, pile height, density, stitches per millimetre, backing and adhesive requirements.[15] Brown[16] has shown that it is easier to remove HDM allergens from synthetic 'flock' carpets than from other types. HDM have a physiological mechanism (suckers on their legs) which allows them to effectively cling to fibres during vacuuming.

It should also be noted that although high performance vacuums incorporating HEPA filters can remove a significant proportion of the HDM allergen in the dust reservoir, they do not remove mites or denature the allergenic proteins. The mite colony thus remains in situ and allergen levels can – given optimum hygro-thermal conditions (80% RH at 25 °C) – rise by geometric progression (times 60 every 30 days).

Steam cleaning at high pressure (6 bar) has been shown to denature the HDM allergenic proteins with an average reduction of c. 92–95% achievable.[17] Studies by Van Strien et al.[18] and Luczynsk et al.[19] found that the concentration of Der pI increases with the age of the carpet. Arlian[20] found that long pile carpets contained significantly more mites than short pile carpets and there was no correlation between mite numbers and frequency or intensity of vacuuming.

Another study by Lewis and Breyesse[21] investigated the retention properties of 26 different carpet types by spiking them all with reference dust containing cat allergens. They were then vacuumed for an equal length of time. They found that the carpet properties most likely to release allergens by vacuuming are: low pile density and height, fluorocarbon coating of fibres, high dernier per filament and low surface area fibres. A study by Price, Pollock and Little[22] showed that woollen carpets have significantly higher concentrations of *Der pI* in the air above them than synthetic carpets, possibly due to electrostatic charge.

There appears to be some anomalies with the Braun *et al.*[13] paper previously cited. The computational fluid dynamic (CFD) model predicted no significant variation in PM_2 airborne particle concentrations between hard and carpeted floors, when a person walked through the room for a total time-span of 40 s. For PM_{10} the predicted disturbance to the dust reservoir results in the airborne concentration increasing from zero to just over $50 \, \mu g \, m^{-3}$ for hard floor coverings, after approximately 50 s. The difference between the hard and carpeted flooring is less than $5 \, \mu g \, m^{-3}$. Such a small differential does not support the paper's conclusions, especially as the CFD modelling technique has not yet been experimentally validated. Given this background, and the fact that it was commissioned by the Carpet and Rug Institute of the USA and their suppliers, it is difficult to concur with the paper's bold assertions that, 'established wisdom regarding the adverse role of carpets on indoor air quality is simply wrong... and that carpeting has the potential to actively contribute to maintaining and improving indoor air quality'.

If this CFD model does turn out to be a close facsimile of what occurs in the field, it appears that as little as 40 s of walking at $4 \, ft \, s^{-1}$ can result in a significant level of sedentary particles being re-suspended to a height of three (child height) and six feet (adult height). This will result in children being likely to inhale significantly more disturbed PM_{10} particulates than adults, as the concentration is greater, *c.* $20 \, \mu g \, m^{-3}$, at this height.

A comprehensive study of flooring by Cole *et al.*[23] in 1996, reported that flooring surface pollutants can contribute to and be reflective of, airborne levels, and airborne levels can contribute to and be reflective of, surface contamination. Dybendal and Elsayed[24] indicated that carpeted floors accumulate significantly more dust, HDM proteins and allergens per unit area, when compared with smooth floors. Furthermore, in a study by Shaughnessy *et al.*[15] that compared both tiled and carpeted floors in American schools, both the airborne mean particle counts and the grams of dirt per square metre were significantly higher for carpeted classrooms in the same school. Of the ten comparable classrooms the figures for grams of dirt per square metre on the tiled and carpeted floors were: 0.015 and 50, 0.21 and 29, 0.14 and 7.7, 0.096 and 23 and 0.055 and 99. There was, on average, 0.103 g weight of dirt per square metre on the tiled floors. This compared with an average of 41.74 g of dirt per square metre on the carpeted floors; over 400 times the weight. Mean particle concentrations ($PM_{2.5}$ and $PM_{>5}$) were also taken for these classrooms and the external ambient air. With an average background ambient air concentration of $2.77 PM_{>5} m^{-3}$ and $10.8 PM_{2.5} m^{-3}$ the comparison between air counts was now much closer at $5.94 PM_{>5} m^{-3}$ and $11 PM_{2.5} m^{-3}$ for tiled floors versus $8.34 PM_{>5} m^{-3}$ and $18.2 PM_{2.5} m^{-3}$ for carpets. This appears to confirm that although the tiled floors had relatively little contamination, any dust was easily disturbed and re-suspended. The claim that carpets act to retain dirt and reduce airborne particulate matter cannot be supported. These classrooms (which were the subject of a regular cleaning programme) had significantly higher airborne particulate counts. Indeed it appears that for $PM_{2.5}$ the carpets are acting as a significant reservoir, boosting internal airborne burdens to a level well above ambient airborne concentrations.

Again the literature would support the view that, although dust particles on hard floor surfaces are relatively easy to disturb, the floors themselves are far cleaner and thus the considerably smaller reservoir of dust leads directly to lower airborne particle counts. There is little or no compelling evidence to confirm – or even support the view – that carpets act as effective sinks, locking in dirt and PM. All the studies appear to confirm that carpets attract dirt, are harder to clean and this results directly in higher airborne particle counts when the surface is disturbed.

Furthermore, Warner et al.[25] have shown that children in the first few years of life – when the immune system is relatively immature – are particularly prone to dust mite sensitisation. The early years of childhood are typified by near-carpet activities. It may not be surprising that in the last 30 years the prevalence of childhood asthma has risen to one in seven. Warner[26] also carried out a major literature review of 182 papers relevant to the area of allergic diseases and the indoor environment. The conclusions appear to concur with the hypothesis under examination:

(i) Up to 100,000 mites may live in $1 m^2$ of carpet.
(ii) Mite populations and their allergens are usually found at their highest level per unit weight of dust in beds, but carpets can contain the largest reservoir in total amount of mite allergens in the house.
(iii) The presence of fitted carpets is particularly strongly associated with high levels of pet allergens.
(iv) The best results for reducing exposure to HDM allergens have been achieved with a combination of encasing bedding and removing carpets and soft furnishings. In all studies where there was a significant benefit to allergy sufferers, carpets were either removed or subjected to intensive treatment.
(v) The UK has the highest consumption of carpets in Western Europe and North America at $3.9 m^2$ per person, with 98% of British homes having fitted carpets compared to 16% in France and 2% in Italy.
(vi) The UK has the highest prevalence of asthma symptoms in 13–14 year olds in the world (19.8%). This is 20 times the rate found in Indonesia.[27]

Warner[28] in challenge tests to inhaled allergens demonstrated that they induce similar reactions to those experienced in a natural allergic reaction. This seems to indicate that airway disease is triggered by inhaled allergen. Air sampling is therefore of importance for indicating levels of allergen that are likely to be inhaled by the patient, although the particle size will influence suspension duration. Air sampling is generally carried out by the use of a personal sampler (sampling the immediate environment that the patient is exposed to) or by a fixed sampler (samples the air in an area where the patient spends most time). Price *et al.*[23] in one of the few studies to investigate indoor airborne allergens, suggest that fixed sampling may be less reliable as the larger units used cleaned the air faster than the allergen was produced. A method has thus to be developed to investigate this possible relationship. Several problems have to be addressed. Firstly, the existing average allergen concentration in the dust reservoir requires to be quantified without significantly reducing the total volume available for possible airborne distribution. Secondly, a known air volume requires to be extracted and the fine dust stored over a period to provide a large enough sample for immunoassay. Thirdly, no unusual circumstances should occur during the testing period that deviate from the normal background activity regimes. It may be a more appropriate strategy for the air sampling pump to extract a similar quantity of air that would normally be respired by one person – over a set period (one or two days). Until a reliable method can be developed for measuring allergen inhalation, present techniques require to be viewed as an index, rather than a measurement, of exposure. It is thus important to agree a methodology to ensure that the results from various studies are directly comparable.

Clinical testing

The health outcomes reported in Chapter 7, were based entirely on self-reported and unsupervised morning and evening peak-flow readings. The veracity of this information is open to some doubt, as several individuals did not complete the protocol. This has to be expected when research is being undertaken with real people in an uncontrolled environment.

A second problem was encountered with the monitoring of drug use. Entry into the study was based on all individuals having been prescribed inhaled drugs – normally steroid based preventers and on-demand dilators. The study team was not competent to offer advice on changes to drug regimes. In general the steroid based anti-inflammatory inhaler (Beclomethasone or similar) was taken in prescribed doses twice daily (first thing in the morning and last thing at night). It is thus relatively easy to measure drug use as each inhaler will last a set and predictable period. The on-demand dilatory inhaler (Salbutymol etc.) is taken by the patient when having an asthmatic incident, to provide immediate symptom relief. As most patients had more than one on-demand inhaler, it was difficult to measure how much had actually been inhaled over any given time frame. Although an attempt was made to prescription profile at source, medical records were generally incomplete and the issuing of prescriptions does not necessarily signify that the doses will all be inhaled. Indeed, most patients were on repeat prescriptions that issued both preventers and dilators in set proportions. Over-prescribing is thus possible. A more accurate and controlled protocol requires to be developed that can measure – at appropriate cyclical junctures – lung function, allergen exposure levels, drug use and general health status.

Cohort selection criteria

Due to the cohort being heavily skewed towards children under 16 – whose asthma can be notoriously capricious – a further confounding variable can be reduced in influence, if the cohort can be selected from only adult asthmatics (over 16 years) with a known history of severe symptoms. Diagnosis of asthma can be established symptomatically by confirmation of episodic wheezing, chest tightness and/or dyspnea and objectively confirmed by evidence of variable airflow obstruction with an increase in FEV1 (forced expiry volume) of $\geq 12\%$ after inhaling Salbutamol ($400\,\mu g$) or variable peak-flow readings of $\geq 15\%$. Forced expiratory volume is not only a test of lung capacity – normally undertaken by the more simplistic peak-flow meter – but can measure the speed at which a patient can inflate and deflate the lungs. Such characteristics can provide additional evidence of the individual's asthmatic state to pulmonary physicians. Patients should also be in receipt of inhaled steroid treatment and have daily symptoms. As the intervention is not designed to assist non-HDM sensitive asthmatics, all volunteers should be skin prick tested to confirm a minimum immunological reaction to HDM proteins.

Clinical testing

It is important to test patients on a regular cyclical basis and accurately measure a greater range of health states. Diary cards recording daily asthma symptoms, peak-flow and inhaled Beta$_2$-agonist use, can provide longitudinal baseline measurements. Other life events that are important to record are: emergency/out of hours visit of patients to the GP, GP visits to patients at home, and A and E hospital attendance. Spirometric measurements can be made before and after the inhalation of a measured amount of salbutamol. This involves the patient blowing into a contraption, which can measure volume and air speed, and subsequent computer analysis can benchmark lung function to known performance targets, given a patient's size, age, sex and weight. If these measurements can be performed at the same time of day (am or pm) for each patient, a detailed profile can thus be assembled.

As antibodies to the specific HDM antigen can be detected in blood, any drop in exposure levels through time will result in a reduction in IgE. Blood samples can be drawn at each monitoring cycle to test for longitudinal changes before and after any intervention.

The Euroquol[29] questionnaire can be used to measure patients' self-reported health status, which will allow comparison with other similar interventionist approaches. In addition the use of the St George's Respiratory Questionnaire[30] can also be used to help measure asthma specific and generic measurements of quality of life. Any improvements in lung function can trigger a phased drug step-down regime. If lung function can be maintained and/or improved with a significant reduction in steroid or dilator use, the economic benefits and cost-effectiveness of such an approach can be maximised.

Economic outcomes

Any attempt to evaluate the cost-effectiveness of the intervention, measuring capital cost against the reduction in drug use and primary health care costs, requires an accurate and reliable regime detailing prescription dispension and inhaler use. Although the two illustrative case studies, outlined in Chapter 7, did show that the intervention had a simple pay-back period of between 26 and 27 months, the calculation was based on the reduction in GP visits and the number of issued prescriptions. There was no attempt to step-down the drug use or monitor unused inhaler doses. There was also no separate assessment made of wider socio-economic factors such as lost work/school days and the information on acute hospital incidents was not recorded.

A study by Cunningham et al.[31] in the United States, evaluating a range of asthma management programmes (AMPs), claimed to demonstrate cost savings of between 55 and 75% were possible, with the highest savings likely to be achieved for young children. From ten studies they derived the unit costs for drug use, GP visits, hospital stays and emergency room treatment per capita. On average, a moderate to severe US asthmatic attends his or her GP 4.65 times a year at a unit cost of $129 per visit ($600). Hospitalisation costs were calculated to be $2 867, with an average of 0.95 days per person per annum, producing a figure of $2 713. Emergency room visits have a unit cost of $285 with an average of 3.48 per annum ($990). The total drug cost for children under five was $558 per annum, with children over five and adults costed at $762. With over 17.3 million Americans afflicted with the condition, including five million children,[32] asthma can be considered an expensive disease and it is the medical insurance companies that are driving these AMPs, which focus on the avoidance of five environmental triggers: environmental tobacco smoke, dust mites, pets, cockroach allergens and mould.

If such programmes could be implemented in Britain, where there is estimated to be 3.864 million GP visits per year and 20 000 acute hospital admissions,[33] the savings per annum could be in the region of £467.5 to £637.5 million. When other associated costs are factored in, such as lost productivity and additional social security payments, the total annual cost is estimated to be £2.237 billion.[33] It is thus imperative that research is undertaken to quantify the likely scale of savings, which could justify a switch from acute and primary care to a more preventative strategy.

The essence of economic appraisal is to compare and value the outcomes of a project with its inputs. In its broadest sense this means comparing benefits with costs. Future research should thus be based on randomised control trials with two arms, where active ventilation is compared with 'control' ventilation. The key hypothesis is that denaturing and allergen avoidance techniques, in conjunction with active ventilation, 'works' i.e. it is predicted to give outcomes significantly different from those of the control group.

From the economic point of view it is thus necessary to evaluate the costs and benefits of both arms, and to compare the results. This involves three steps: identification of all relevant categories of outcome; measurement of all these categories; and valuation of all these categories. If valuation can be accomplished, then the appraisal of the project boils down to a comparison of the net benefits (benefits minus costs) of each arm. The aim must be to judge the incremental net benefits of any active intervention over the control. If full 'monetisation' is not possible, then cost per 'quality-adjusted life years' (QALYs) can be calculated using results from the Euroquol instrument. It is also possible to calculate 'cost-effectiveness' in terms of cost per given change on the St George's Respiratory questionnaire.

In considering the impact of the disease, two concepts can be combined which incorporate both premature mortality and years lost to disability. The GINA[34] study calculated that worldwide, asthma has an average score of 15 'disability adjusted life years' (DALYs). Of course in countries such as Scotland – that are at the top end of the prevalence scale – this figure is likely to underestimate the years lost to disability.

Air infiltration rates

As previously outlined in Chapter 5, a dwelling's volume and background air infiltration rate will play a crucial role in determining water vapour diffusion rates. Such may also influence the efficacy of any retrofit mechanical ventilation regime. If the underlying hypothesis is correct – that HDM colonisation and proliferation rates are positively correlated with high internal RH – dwellings that incorporate flues or have draughty windows or 'loose' construction techniques, are unlikely to suffer from high levels of dust mite activity. They are, however, likely to be cold, uncomfortable and/or energy profligate. The challenge is thus to strike the optimum balance between internal comfort, energy efficiency and 'healthy' ventilation rates. This cannot be done without at least scoping the characteristics and likely variation in some generic house types. Chapter 5 attempted to model and norm reference five of the main Scottish public-sector house types. This required many assumptions to be made, such as crack length and size around windows, and may not be an accurate facsimile of reality. Attempting to produce a definitive answer to this question would require a large number of dwellings to be tested using 'blow door' or gas tracer techniques which will allow a more accurate assessment of planned and unplanned air infiltration. This of course applies mainly to existing dwellings, built under regulations which paid little heed to air tightness. A range of techniques can be adopted to ensure that new dwellings are 'tight' ($c.$ 3–5 ach^{-1} at 50 Pa pressure differential). This will allow mechanical ventilation with heat recovery to be more effective in extracting moisture and air pollutants from dwellings, while simultaneously delivering fresh, filtered pre-warmed air to inhabited rooms.

References

[1] Macher JM, Tsai FC, Burton LE and Lui KS. Concentrations of cat and dust mite allergens in 93 US office buildings. The International Academy of Indoor Air Sciences, Monterey. *Indoor Air*, July 2002, Vol. IV, pp. 359–364.

[2] Lewis RD, Breysse PN, Lees PS, Deiner West M, Hamilton RG, Eggleston P. Factors affecting the retention of dust mite allergen on carpet. *American Industrial Hygiene Journal*, 1998, Vol. 59, pp. 606–613.

[3] Dingle P and White K. Comparison of PM$_{10}$ concentration in carpeted and non-carpeted residential dwellings. February 2000.

[4] Jorde W. Carpets and allergenic disease. International man made fibres Conference, September 1993.

[5] Voute PD, Zock, JP, Brunkenreef B and de Jongste JC. Peak-flow variability in asthmatic children is not related to wall-to-wall carpeting on classroom floors. *Allergy*, 1994, Vol. 49, pp. 724–729.

[6] Sidenius KE, Hallas TE, Brygget T, Poulsen LK and Mosbech H. House dust mites and their allergens at selected locations in the homes of house dust-mite allergic patients. *Clinical and Experimental Allergy*, 2002, Vol. 32, pp. 1299–1304.

[7] Platts-Mills TA, Thomas WR, Aalberse RC, Vervlet D and Chapman MD. Dust mite allergens and asthma: report of a second international workshop. *Journal of Clinical Immunology*, 1992, Vol. 89, pp. 1046–1060.

[8] Platts-Mills TAE, Ward GW, Sporik R, Gelber LE, Chapman MD and Heymann PW. Epidemiology of the relationship between exposure to indoor allergens and asthma. *International Archives of Allergy and Applied Immunology*, 1991, Vol. 94, pp. 339–345.

[9] White K and Dingle P. The effect of intensive vacuuming on indoor PM mass concentration did find that intensive vacuuming. The International Academy of Indoor Air Sciences, Monterey. *Indoor Air*, July 2002, Vol. III, pp. 92–97.

[10] Kemp PC, Dingle PW and Neumeister HG. Particulate matter intervention study: A causal factor of building related symptoms in older buildings. *Indoor Air*, 1998, Vol. 3, pp. 153–177.

[11] Mitakakis TZ, Mahmic A and Tovey ER. Comparison of vacuuming procedures for reservoir dust mite allergen on carpeted floors. Environmental and occupational disorders. *Journal of Clinical Immunology*, January 2002, pp. 122–124.

[12] Braun WX, Cicciarelli BA, Davidson DL, Hart EH, Luedtke A, McIntosh K and Peoples PR. Indoor pollutant measurement and modelling comparing impact of surface characteristics. The International Academy of Indoor Air Sciences, Monterey. *Indoor Air*, July 2002, Vol. I, pp. 885–890.

[13] Brown SK. Application of an optimised procedure for measuring house dust mite numbers in carpet and furniture. *Indoor Air: An integrated approach*, Oxford, UK, Elsevier Science, 1995, pp. 63–66.

[14] Gunnerson L, Sidenius K and Hallas TE. House dust mites, humidity of room air and flooring materials. The International Academy of Indoor Air Sciences, Monterey. *Indoor Air*, July 2002, Vol. IV, pp. 725–730.

[15] Shaughnessy RJ, Turk B, Evans S *et al*. Preliminary study of flooring in the US: airborne particulate exposures in carpeted vs uncarpeted classrooms. The International Academy of Indoor Air Sciences, Monterey. *Indoor Air*, July 2002, Vol. I, pp. 974–979.

[16] Brown SK. Two types of carpet with reduced house dust mite prevalence. The International Academy of Indoor Air Sciences, Monterey. *Indoor Air*, July 2002, Vol. I, pp. 1026–1031.

[17] Howieson SG, Lawson A, McSharry C, Morris G, McKenzie E. Indoor air quality, dust mite allergens and asthma. The International Academy of Indoor Air Sciences, Monterey. *Indoor Air*, July 2002, Vol. I, pp. 113–118.

[18] Van Strien RT, Verhoff AP, Brunekreef B and Van Wijnen. House dust mite allergen and respiratory symptoms in children: a case controlled study. *Clinical Experimental Allergy*, 1994, Vol. 24(11), pp. 1061–1069.

[19] Luczynska C, Sterne J, Bond J, Azima H and Burney P. Indoor factors associated with concentrations of house dust mite allergen, Der p1 in a random sample of houses in Norwich, UK. *Clinical Experimental Allergy*, 1998, Vol. 28, pp. 1201–1209.

[20] Arlian LG, Bernstein IL and Gallagher JS. The prevalence of house dust mites and associated environmental conditions in homes in Ohio. *Journal of Allergy and Clinical Immunology*, 1982, Vol. 69, pp. 527–532.

[21] Lewis RD, Breyesse PN. Carpet properties that affect the retention of cat allergen. *Annals of Allergy, Asthma & Immunology*, 2000, Vol. 84, pp. 31–36.

[22] Price JA, Pollock I, Little SA, Longbottom JL and Warner JO Measurement of airborne mite antigen in homes of asthmatic children. *The Lancet*, 1990, Vol. 336, pp. 895–897.

[23] Cole EC, Dulaney PD, Lesse KE, Foarde KK, Franke DL, Myers EM and Berry MA Bio-pollutant sampling and analysis of indoor surface dusts: Characterisation of potential sources and sinks. *American Society of Testing and Materials*, Standard Technical Publication, 1996, Vol. 1287, pp. 153–165.

[24] Dybenal T and Elsayed S. Dust from carpeted and smooth floors: VI Allergens in homes compared with those in schools in Norway. *Allergy*, 1994, Vol. 49(4), pp. 210–216.

[25] Warner JA, Little SA, Pollock I *et al*. The influence of exposure to house dust mite, cat, pollen and fungal allergens in the home on primary sensitisation in asthma. *Paediatric Allergy Immunology*, 1990, Vol. 1, pp. 79–86.

[26] Warner J. Allergic diseases and The Indoor Environment, report commissioned by the Healthy Flooring Network, http://www.healthyflooring.org/Allergic%20 Diseases.pdf.

[27] ISAAC Worldwide variation in prevalence of asthma, allergic rhinoconjunctivitis, and atopic eczema: ISAAC. *The Lancet*, 1998, Vol. 351, April 25, pp. 1225–1232.

[28] Warner JO. Significance of late reactions after bronchial challenge with house dust mite. *Archives of Disease in Childhood*, 1976, Vol. 51, pp. 905–911.

[29] Dorman PJ, Slattery J, Farrell B, Dennis MS and Sandercock PAG. A randomised comparison of the Euroqol and Short Form-36 after stroke. *British Medical Journal*, 1997, Vol. 315, p. 461.

[30] McMaster University. Asthma quality-of-life questionnaire. Department of Clinical Epidemiology and Biostatics, McMaster University Medical Centre, 1200 Main Street West, Hamilton, Ontario, Canada, L8N 3Z5.

[31] Cunningham K, Mudarri D, Heil M and Apelberg B. Medical cost savings obtained through asthma management programs and environmental trigger avoidance. The International Academy of Indoor Air Sciences, Monterey. *Indoor Air*, July 2002, Vol. I, pp. 784–789.

[32] US EPA, 2000. The EPA children's environmental health yearbook supplement. EPA-100-R-00–018, August 2000.

[33] National Asthma Campaign, Out in the open, A true picture of asthma in the United Kingdom today. *The Asthma Journal*, September 2001, Vol. 6, No. 3.

[34] Masoli M, Fabian D, Holt S and Beasley R. Global Burden of Asthma, Global Initiative for Asthma, Medical Research Institute of New Zealand and University of Southampton, 1994, Summary report (www.ginasthma.com) quoting the WHO, World Health Report 2002 – Reducing risks, promoting health life.

Chapter 9

Hazardous indoor pollutants

Regulatory guidelines and hazardous indoor pollutants

In 1991 the House of Commons Select Committee established to investigate indoor air pollution concluded that 'Overall there appears to be a worryingly large number of health problems connected with indoor pollution which affect a large number of people'.

Thirteen years on, the UK is only now considering issuing some guidance on five compounds (nitrogen dioxide, carbon monoxide, formaldehyde, benzene and poly-cyclic aromatic hydrocarbons (PAH). Guidance, however, does not constitute a standard, and it remains unlikely that regulations aimed at ensuring good air quality in residential buildings will be forthcoming.

At the outset of the twentieth century there were approximately 50 materials used to construct buildings. By the end of the century Raw[1] claimed that this list had grown to around 55000, with half of them being synthetic. Compounds found in indoor air may have off-gassed from the building materials, furnishings and fittings, internal processes, cleaning products, with – somewhat ironically – even air fresheners, being implicated in contribution to indoor air pollution. Such pollutants can be grouped under the following alphabetical classifications: ammonia, asbestos, benzenes, biocides, carbon dioxide, carbon monoxide, detergent, dust, ethanol, fibreglass, formaldehyde, hydrocarbons, hydrogen chloride, methanol, microorganisms, motor vehicle exhaust, nitrogen oxides, ozone, paint, poly-chlorinated biphenols (PCB), pesticides, photochemical smog, radon, solvents, sterilant gases, sulphur oxides, tobacco smoke and vinyl chloride.

The most common gases found in the indoor environment are carbon dioxide/monoxide, nitrogen and sulphur dioxide, VOCs, radon, formaldehyde and ozone. The most common suspended PM (both solid and liquid) are asbestos fibres, fibrous particulates (fibreglass or rockwool), bacteria and fungi, tobacco smoke, HDM allergens, pollen and dust.

A recent study by the World Wide Fund for Nature[2] (WWF), which tested for 77 synthetic chemicals in the blood of 150 volunteers, identified a median of 27 per person. Many of these chemicals are routinely found in food (organochlorine pesticides) and the domestic environment (PCBs in PVCu and PBDs used as flame retardents in furniture) and are considered to be persistent bio-accumulative endocrine disrupters. With over 100 000 synthetics available for inhalation or ingestion these tests clearly represent the tip of a toxic chemical iceberg. As most bio-medical studies at best attempt to quantify the risks of exposure to a single compound or group (PCB's and DDT have been found to act as false oestrogens compromising male fertility), a study that incorporates synergistic, additive or antagonistic chemical and biological reactions, would have to address an unmanageable number of variables. Complexity and difficulty do not, however, provide an excuse for doing nothing in terms of research, guidance and the development of legally binding standards. Should the onus not be put on the manufacturers to demonstrate safety (both singularly and in combination) rather than on society being used as a living test laboratory to eventually quantify the level of hazard?

Sick building syndrome

When a building is subject to complaints sufficient to convince management to conduct an indoor air quality monitoring exercise it will normally be classified as suffering from 'sick building syndrome' (SBS). In the early 1980s, the WHO[3] attempted to define this phenomenon on the basis of a group of frequently reported symptoms (sensory irritation in the eyes, nose and throat; neurotoxic or general health problems; skin irritation; non-specific hyper-sensitivity such as pneumonitis, humidifier fever, asthma and rhinitis and noxious odour and taste sensations), without a specific aetiological causal factor being identified. Although there is undoubtedly a psychosocial element to workplace satisfaction/dissatisfaction, when similar groups have been compared, those working in buildings diagnosed as 'sick' invariably have physical, biological or chemical components in the mix. The WHO characterisation of SBS is based on the theory that most complaints of a sensory nature are a consequence of the non-specific irritation or overstimulation of trigeminal nerves in mucous membranes. These nerves respond to chemical odours that cause irritation, tickling or burning. This is a defence mechanism triggering mucous production, sneezing or interruption of breathing, to protect the body from what are perceived to be potentially harmful substances.

The effects of SBS, and exposure to hazardous indoor air pollutants, are not restricted to the respiratory tract. At the psychosocial level a study by Morrow et al.[4] on 22 workers in Minnesota – who were regularly exposed to organic solvents – revealed a high level of somatic disturbance, anxiety, depression, poor concentration, social isolation and fear of losing control.

This investigation is of course significant, in that it discovered that exposure to certain chemicals caused changes in neurological behaviour, which would normally be diagnosed as psychosomatic in nature. A putative mechanism for multiple chemical sensitivity (MCS) has been proposed by Ashford and Miller[5] and Bell *et al.*,[6] and appears to involve a similar dose–response relationship for allergen sensitisation. They suggest that it is the olfactory-limbic pathways in the brain that are affected by exposures to solvents and pesticides. The limbic system, which includes the hypothalamus, governs the interaction between an individual and their environment. It has been described as the primitive 'smell brain' and controls a variety of human emotions associated with self-preservation such as food finding, eating, fighting and personal protection. The olfactory bulbs are in close proximity to the limbic area and supply much of the neural input. Olfactory nerves link the limbic system to the external chemical environment. Since there is no blood–brain barrier in the limbic region, various substances can enter the olfactory bulbs and be transported in the neurons. Isolating the particular substance or combinations, which can drive a plethora of symptoms – some immediate and some delayed – requires highly sophisticated diagnostic techniques.

SBS thus appears to be applied to instances where there is a diagnostic failure to isolate the main compounds responsible, individually or in synergistic combination. SBS can, in the main, be attributed to such mundane factors as insufficient ventilation provision or hygro-thermal control; inadequate maintenance of plant, filters and air distribution ducting; changes in contaminant loads; changes in operational norms and poor design strategies or plant commissioning.

This is important in western society where most of the time is spent indoors. In developing countries there are still communities where only a small part of the day is spent indoors, and in each of these, the prevalence of wheezing among children remains low. By contrast, in urban communities in developing countries and even more so in American/European cities, the population is increasingly indoors and asthma has become an epidemic disease that disrupts many families and causes large numbers of hospital admissions.[7] This move indoors has inevitably involved major changes in the air we inhale, including effects of protein allergens, bacterial endotoxins, hydrocarbons and gases such as ozone and NO_2.

The respiratory tract is a major entry route and end organ for the effects of indoor pollutants. Understanding of the effects of various irritants is complicated, not only by complex mixing, but additionally by host response variability. Inhaling a wide range of diverse materials – even at very low concentrations (ozone, for example, can cause lung inflammation at 0.08–0.12 ppm[8]) – at a rate of 10,000–20,000 litres per day[9] can produce both immediate and delayed respiratory complaints. Bascom *et al.*[8] classified indoor pollutants by their effects on human tissue; corrosives are so termed because they cause direct destruction of structural tissue; irritants are compounds that cause symptoms and inflammation (i.e. trigger an influx of cells and mediators); sensitisers are materials that induce a specific immunologic response (HDM allergens triggering IgE) with the well-known properties of specificity, diversity and amplification.

The exacerbation and development of asthma caused by exposures to specific airborne compounds such as ozone, carbon monoxide, oxides of nitrogen and sulphur, lead and various PM, are routinely addressed by the US Environmental Protection Agency in setting national air quality standards. Possible correlations between exposures to numerous compounds, categorised as air toxics – and the exacerbation of asthma, remain seriously under-researched. Pollutants such as diesel particulates, ETS, VOCs, metals, pesticides and endotoxins, have all been implicated in respiratory disease, at least as symptom triggers, if not as primary sensitisers. This chapter will review some of the major indoor gases and compounds implicated in either pre-disposing the immune system or triggering asthmatic symptoms.

Carbon dioxide

Carbon dioxide has generally been used as a marker for ventilation rates and a proxy for indoor air quality. The upper limit for CO_2 concentration level as recommended by ASHRAE[10] is 700 ppm. Nocturnal build-ups in small bedrooms where the construction is relatively 'tight' are common. It has been suggested by Carrer and Maroni[11] that the increase in morbidity and mortality for asthma and allergies may be, in part, due to an increase in exposure to indoor air pollutants. Kim et al.[12] using multivariate analysis in a Korean study claimed that indoor CO_2 concentrations were associated with wheezing attacks in children with a history of asthma. The mean value for CO_2 concentrations in these dwellings was 564 ppm. The study did not attempt to uncover a mechanism and simply claimed that CO_2 levels may be considered a good surrogate for concentrations of other indoor pollutants.

The reduction in oxygen levels associated with a build-up of CO_2 will require subjects to increase their rate of respiration, particularly if they are engaged in physical activity. There is some evidence to show that reduced oxygen levels will cause headache and nausea. In extreme cases subjects may see interference patterns and lose consciousness. Increased respiration rates may also result in the inhalation of larger airborne allergen burdens – another hypothesis that requires to be tested.

A study in Texas by Corsi et al.[13] investigating CO_2 levels in elementary schools found that 88% of 120 randomly selected classrooms had peak levels above 1000 ppm, and 21% had peak levels above 3000 ppm. Levels of CO_2 greater than 1000 ppm (or 700 ppm above outdoor background) are generally assumed to be indicative of inadequate ventilation. A similar study by Sowa[14] looked at CO_2 concentrations in 28 classrooms in Warsaw. Measured levels during the working week when the classrooms were occupied, were always above 1000 ppm and peaked at 4000 ppm (20% cumulative frequency). In addition to CO_2 monitoring, tests were undertaken to detect VOCs (such as formaldehyde, toluene, xylenes, decane, undecane, pentane, acetone, heptane, hexane and isopropyl benzene). In 5 out of 27 cases these chemicals were found to be above the permitted concentrations. As many of these gases are toxic and implicated in triggering asthmatic symptoms, it seems reasonable to use CO_2 as a proxy for general indoor air quality and ventilation rates.

In a study of an office building in Beijing by Wang and Zhao[15] where 98.7% of occupants reported one or more physiological symptoms considered to be signifiers for SBS, afternoon CO_2 concentrations of between 700 and 1000 ppm were measured. Such a level was associated with the build-up of a range of VOCs that are associated with SBS. The case for using CO_2 as a marker for air quality has also been made by Health Canada.[16] It would thus be prudent to include in any further trials, at least a scoping study to measure any nocturnal build-up of carbon dioxide levels in bedrooms, not only as a measure of ventilation rates but as an indicator of possible associated pollutant build-up. Carbon dioxide can also be used to test air infiltration rates if a set quantity is released and the decay rate monitored until the preceding ambient level is reached.

Carbon monoxide and nitrogen dioxide

Although the prevalence of asthma in the UK is fairly equally distributed geographically, there are more asthmatic attacks in urban areas, particularly when pollution levels rise. There are three main sources of carbon monoxide (CO) that can be found in the internal environment – traffic fumes, gas appliances and cigarette smoking. Carbon monoxide is a toxic gas which mixes with the haemoglobin in the bloodstream and at high concentrations will result in death. The WHO[17] set a guideline value of $10 \, mg \, m^{-3}$ for any 8-hr period in the home. A recent study by Raw et al.[18] measured CO levels in 876 randomly identified English dwellings, over a period of one year. It found that CO concentrations in kitchens and bedrooms almost doubled during autumn and winter, possibly due to increased heating system use and a reduction in natural ventilation rates. Higher levels were also found in urban areas, in homes with a gas oven, homes with unflued fossil fuel heaters and in the homes of smokers. The CO concentrations on no occasion exceeded the WHO guidelines with the highest geometric mean recorded at $0.62 \, mg \, m^{-3}$ and a maximum concentration of $4.45 \, mg \, m^{-3}$. Nitrogen dioxide (NO_2) levels, however, were above the exposure guideline of $40 \, \mu g \, m^{-3}$ in 25% of all kitchens. This was attributed to gas cookers and the report postulated that the majority of such homes will exceed the WHO air quality guideline for annual exposure to NO_2. The report claimed that indoor exposures mainly arise from indoor sources and will vary seasonally, largely due to ventilation rates and occupant behaviour. Although CO levels in homes were found to be low – particularly over short time spans when gas ovens are in use – NO_2 levels were above the WHO recommendations and could therefore have significant health implications.

Environmental tobacco smoke

Asthma is an immunologic disease triggered by specific allergens as well as respiratory irritants. The effect is one of airway inflammation and accompanying bronchial hyper-responsiveness. A significant relationship between household environmental tobacco smoke (ETS) exposure and childhood asthma has been reported by Morgan and Martiney.[19] Furthermore, a meta-analysis of 73 studies by Vork, Broadwin and Lipsett[20] claimed that the results indicated a strong and consistent association between exposure to ETS and the incidence of childhood asthmatic symptoms. There is however, some doubt as to whether ETS is a sensitiser or simply an irritant, as it does not stimulate marker 'T' cell activity, typical of allergy. Tobacco smoke contains a large number of chemicals considered by the US federal government as hazardous air pollutants (HAPs). In addition to other addictive substances such as nicotine, the OEHHA[21] calculated that one cigarette contains the following weight of hazardous air pollutants: 2267 µg acetaldehyde, 1028 µg acetonitrile, 979 µg formaldehyde, 879 µg toluene, 560 µg acrolein, 461 µg butadine, 417 µg benzene, 300 µg butanone, 181 µg acrylonitrile, 169 µg styrene, 162 µg phenol, 136 µg ethylbenzene, 70 µg cresol isomers, 52 µg methylnaphthalenes, 42 µg naphthalene. The task of measuring such a range of chemicals and their effects on health is clearly complex, however, cigarettes do contain a chemical called 'cotinine' which can be easily detected and measured in saliva and urine. Cotinine levels can thus provide a good proxy for ETS exposure, particularly in children.

   Although ETS and its constituent chemicals are found in various concentration densities per square metre in outdoor air, the close proximity between smokers and non-smokers and the persistence of pollutants in indoor spaces – due to absorption by indoor surfaces – produces an ETS inhalation intake by passive smokers, c. 100 times greater than for typical outdoor concentrations.[22,23] Layton[24] calculated that the average person breathes approximately 12 m^3 of air per day. Nazaroff and Singer[25] using the concept of reference exposure levels (REL) – the background exposure level of the non-smoking population – calculated that passive smokers will inhale 1.9 µg m^{-3} of acrolein in a smokey environment, which corresponds to 30 times the REL of the non-smoking population. Relatively high-risk levels were also found for formaldehyde and acetaldehyde at approximately double the REL.

Although the short- and long-term effects of such high exposure levels are still to be assessed, it is clear that ventilation rates in dwellings play a major role in diluting or exacerbating indoor HAP concentrations and their associated toxic or pulmonary health outcomes. CIBSE[26] recommend an air supply rate for 'smokey' environments of between 16 and $36 \, \mathrm{l \, s^{-1}}$/person. In an average four-person family living room with a volume of $50 \, \mathrm{m^3}$, this would represent an air change rate of between 4.6 and $10.3 \, \mathrm{ach^{-1}}$. Any intervention that increases ventilation rates is, however, likely to reduce ETS exposure in passive smokers.

Bioaerosols, mould and mycotoxins

With the exception of specific chemical and toxic gas incidents, building-related illnesses are usually associated with exposure to bioaerosols.[27] Mould fungal spores have been implicated in many studies[28–31] as a causal factor in lung disease. Although the type and species will vary depending on location, hygro-thermal environment, construction techniques and internal finishes, there are four main types of mould in northern temperate latitudes that are highly prevalent: *Penicillium*, *Aspergillus*, *Cladosporium* and yeasts.

Microbes and bioaerosols have the capacity to cause four types of ill health: allergy, infection, irritation and toxic poisoning. Exposure to allergens can stimulate specific immunological responses such as rhinitis, asthma and pneumonitis. Infection refers to entry and multiplication of a biological agent in a host body such as inhaled *Aspergillus* invading and multiplying in the bloodstream. Irritation from bioaerosols may result in conjunctivitis, rhinitis and asthma, while mould can produce a variety of glucose polymers found in most cell walls, which are associated with airway inflammation. Moulds can also produce mycotoxins that have been associated with cancer and acute poisoning. Croft, Jarvis and Yatawara[32] reported on one such case in Chicago where the mould species *Stachybotrys aura* (*chartarum*) was shown to be the culprit. Mason *et al.*[33] have shown that those who have been exposed to *Stachybotrys* toxins, in sufficient concentrations, usually suffer from mucous membrane irritation, respiratory and nervous system disorders and immune system dysfunction. Although mould will grow on internal surfaces where condensation regularly forms, it can also grow interstitially producing respirable mycotoxins in the indoor environment without obvious visual markers, which would normally instigate some form of remedial treatment. It is this type of insidious invasion that has triggered many insurance companies to withdraw cover and exclude mould from their policies. The implications of this decision could be profound. Building owners or landlords will now have to cover the direct costs and damages of any claims which may arise where mould is shown to be a significant causal factor in the aetiology of the complaint.

In a study of 330 Finnish children, where some dwellings were found to have airborne penicillium microbe counts of $17\,400\,\mathrm{cfu\,m^{-3}}$ (colony forming units per cubic metre), Hyvarinen et al.[34] found no link between measured winter airborne microbe counts and asthma, however, he did suggest that there was a link between moisture damage and asthma. The size of these microbes appears to determine where they are deposited in the upper respiratory tract. Another clinical trial in Finland[35] studying inflammatory reaction in human and mouse lung macrophages concluded that exposure to fungal spores did not increase the production of inflammatory markers in the studied cell lines, whereas bacteria did cause an increase in inflammatory markers.

A major meta-analysis by Fung and Hughson[36] of 416 articles concluded that the current published human studies demonstrate a clear association between allergy and respiratory symptoms, with exposure to moisture and mould. Mould normally requires a humidity of over 70% to germinate. Micro-climates provided by carpets, bedding and soft furnishings are likely to be significantly damper than the surrounding air. Maintaining the internal RH below 60% should therefore effectively limit the likelihood of microbe viability.

Infectious aerosols are known to contribute substantially to the transmission of such diseases as the common cold, influenza, adenovirus, measles, tuberculosis and other respiratory illnesses. Disease transmission by inhalation of infectious aerosols may be influenced by the following factors: indoor concentration of infectious agents; transport pathways between individuals; viability of infectious agents and the susceptibility of individuals to particular diseases. Building design, construction and system operation can also impact on rates of infectivity. A study undertaken by the US army[37] revealed 50% higher rates of clinically confirmed acute respiratory illness with fever, among recruits in newer barracks with closed windows, low rates of outside air supply and extensive air recirculation, than among recruits in older barracks with frequently opened windows, more outside air and less recirculation. There appear to be seven main factors which could significantly influence infection rates[38]:

1   The rate and effectiveness of outdoor air ventilation which dilutes concentrations of indoor aerosols.
2   The rate and efficiency of air filtration.
3   Disinfection by ultraviolet light which may deactivate infectious organisms.
4   The rate of air recirculation, which influences transport between regions of the building.
5   The occupation density which influences the distance between individuals.

6 The temperature and humidity of air which affects the viability of infectious aerosols and human susceptibility.

7 Indoor toxic fungal exposures which may alter human susceptibility to infection.

The paper goes on to claim that providing 'healthful' air quality in buildings has the potential to prevent five to seven million communicable respiratory infections, saving the US economy between 5 and $75 billion per annum.

Ozone

Studies by Peden, Setzer and Devlin[39] and McConnell et al.[40] have implicated ozone (O_3) as both a sensitiser and a trigger for asthmatic incidents. Monn[41] has claimed that bio-allergens, suspended particles and ozone are the most important pollutants of indoor air. Devlin et al.,[42] using a 'split nose' methodology, demonstrated that O_3 plays a significant role in the exacerbation of respiratory disease, by priming the airway mucosa enhancing the cellular responses to allergens, or by exerting an intrinsic effect on airway inflammation. A study by Bayer et al.[43] monitoring indoor/outdoor O_3 levels every 45 min, demonstrated that, in general, indoor levels remain well below ambient. Internal ozone rose from 9 ppb in the morning to 11 ppb at 18.30, while external ozone – generated by sunlight acting on various pollutants – rose from 11 ppb in the morning to c. 61 ppb in late afternoon. Unless there is an indoor ozone source such as a photocopier, external exposure is likely to be significantly higher. In contrast, the concentration of respirable sized particles was often higher than outdoor concentrations and this suggests that indoor surfaces may have a sink effect, allowing particulate reservoirs to accumulate. Any significant increases in internal air movement and currents can thus drive re-suspension cycles.

Toxic chemicals

Numerous questions exist regarding possible associations between exposures to airborne toxic chemicals and the development or exacerbation of asthma. A meta-analysis of the literature by Barry et al.[44] attempted to evaluate and categorise over 5000 separate papers published since the early 1960s. From this quarry they developed a protocol to extract the most significant components using a five-point (poor to excellent) scale. The result was a distillation of 752 papers separated into 19 categories. These were, in decreasing incidence: ETS (150), overview (148), non-halogented hydrocarbons (90), aldehydes (86), metals (85), diesel exhaust particulates (73), nitrogen containing compounds (43), miscellaneous (33), pesticides and herbicides (30), esters and phalates (27), fibres (16), halogenated hydrocarbons (14), acids (9), amines (9), glycol ethers (3), ketones (2) and nitrosoamines (2). Most of the papers concentrated their efforts on studying one specific variable. Little work had been undertaken on the possible symbiotic effects that may occur when several chemicals sources are present. Is there, for instance, an indoor equivalent of the photochemical reactions that develop during an external smog incident? A study by Brugge et al.[45] concluded that due to the complexity of variables and confounders, a multi-factorial approach has to be adopted and that no single variable or component appears to have a dominant influence.

Volatile organic compounds

Studies by the BRE[46] into indoor VOC sources and personal exposure to VOCs, concluded that of the 24 sources identified in Figure 9.1, indoor exposure was far more significant than the outdoor levels for six of the most common air pollutants (Figure 9.2).

Further studies by Coward et al.[47] looking at VOCs in a range of dwellings, categorised by date of construction, concluded that VOC levels were rising due to the historical reduction in ventilation rates, combined with the type of construction materials now prevalent e.g. chipboard flooring, PVC tiles, paint finishes etc. Figure 9.3 shows this trend through time while Figure 9.4 confirms that levels in the most recent dwellings are now reaching concentrations that must give cause for concern.

Volatile organic compound levels reduce through time as the components release their embodied compounds. There is evidence to show that off-gassing rates can also be affected by humidity and temperature, with concentrations peaking in autumn.

A study of 5951 elementary-schoolchildren in Russia by Jaakkola et al.[48] attempted to relate asthmatic symptoms, persistent cough, respiratory infection and allergy to emissions from surface materials as indicated by recent renovation. Emission rates are typically at their highest over the first few weeks from new materials such as PVC flooring, soft furnishings, particleboard, suspended ceilings, synthetic carpets, wall coverings and painting. They hypothesised that the multiplicity of chemicals emitted from such materials may cause inflammation to the respiratory tract. Although the study was undertaken by self-reported questionnaire, and had a multiplicity of potential confounders such as gender, heating type, ETS, dampness, household numbers and parents' education (which the team tried to account for) they did report a clear association between respiratory symptoms of the lower respiratory tract and the recent presence of materials known to off-gas a wide variety of chemicals. New synthetic carpets provided the most significant correlation. In a separate study, Jaakkola et al.[49] also found that children exposed to PVC flooring in nurseries and bedrooms

Activities
- (A) Cigarette smoking
- (B) Painting windows doors etc.
- (C) Furniture polishing
- (D) Cooking
- (E) Car exhaust/petrol vapour
- (F) Ventilation (outdoor air pollution)

Products
1. Carpet
2. Vinyl flooring
3. Carpet underlay
4. Self leveling screed
5. Liquid applied damp proof membrane
6. Particleboard furniture
7. Wallpaper
8. Walls painted with emulsion
9. Cavity wall insulation
10. Sealant around worktops in kitchen/bathroom
11. Dry cleaned clothes

12. Store of cleaning materials
13. Toiletries in bathroom
14. Timber in joists
15. Curtains
16. Vinyl adhesive
17. Open fire
18. Perfume
19. Chipboard flooring
20. Moth repellant
21. Air freshner
22. Printing ink
23. Pesticides
24. Glues for hobbies

Figure 9.1 Indoor VOC sources.
Source: Crump, BRE[29].

had an 80% higher risk of asthma than those in PVC-free homes. PVC contains a highly toxic range of bi-phenols and other VOCs.

A study by Rumchev *et al.*[50] measured the ten most common VOCs found in the domestic environment: benzene, toluene, *m*-xylene, *op*-xylene, ethylbenzene, styrene, chlorobenzene, 1,3-dichlorobenzene, 1,2-dichlorobenzene and 1,4-dichlorobenzene. When comparing the 88 cases against 104 controls, the total VOC (TVOC) burden was over three times higher in the asthmatics' dwellings ($37.1\,\mu g\,m^{-3}$ compared with $11.7\,\mu g\,m^{-3}$). Dust mite allergen levels were also *c.* six times higher in the asthmatics' dwellings. The mean concentration of $1.99\,\mu g\,gm^{-2}$ – which although nominally below

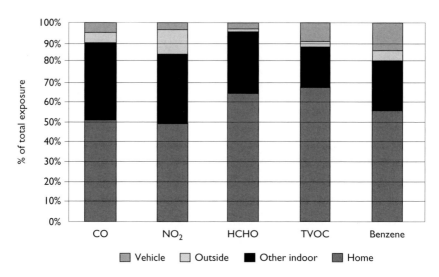

Figure 9.2 Exposure levels of 30 individuals in Herts.
Source: Crump, BRE[40].

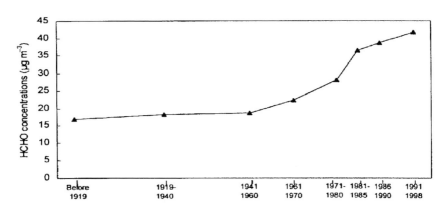

Figure 9.3 Formaldehyde concentrations by dwelling age.
Source: Coward[47].

the WHO sensitisation threshold – produced a positive and statistically significant correlation with dust mite allergens and occupancy levels, with every additional person being associated with an HDM allergen increase of $3 \, \mu g \, g^{-1}$ of fine dust. New carpets in the dwellings increased the TVOCs by an average of $133 \, \mu g \, m^{-3}$ while new furniture produced an average increase of $95 \, \mu g \, m^{-3}$. The presence of smokers increased the levels of particles by

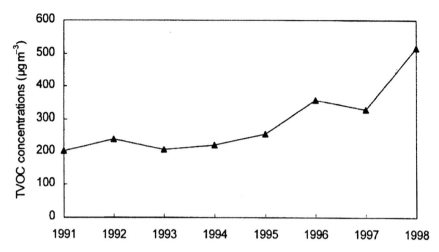

Figure 9.4 TVOC in new dwellings.
Source: Coward[47].

$67 \, \mu g \, m^{-3}$. A logistic regression model was used to explore any interaction between the TVOCs, formaldehyde and dust mite levels and concluded that no significant interaction could be found between these three risk factors. This suggests that TVOCs, formaldehyde and dust mite allergen levels are independent risk factors for asthma. The Coward[47] study into VOCs in 876 dwellings in England had eight pertinent conclusions:

1 Indoor TVOC concentrations vary significantly with season. Off-gassing appears to be affected by a combination of activities, temperature and ventilation.
2 Benzene and toluene were positively correlated with smoking (as they are both constituents of tobacco smoke this is somewhat unsurprising).
3 A built-in garage significantly increased TVOCs with a range of compounds found in paints, petrol and solvents migrating directly into the dwelling.
4 Benzene and *m/p*-xylene were found in higher concentration in metropolitan areas. As they are constituents of petrol fumes this is not surprising. *m/p*-Xylene was also correlated with dwellings containing air fresheners.
5 Undecane was significantly increased by frequency of indoor painting.
6 The main sources of VOCs were: stored materials and car fuel; frequency of decoration and type and age of fittings and furnishings; combustion of methane and ETS.
7 There appeared to be no overall factor representing ventilation rates.

8   Most homes had VOC levels well below existing guidelines apart from benzene which was found in over 50% of dwellings to be above the National Air Quality Guidelines' (NAQS) target for 2010.

There is however a problem with making such simple comparisons as the measured levels were taken as geometric mean values over a consecutive four-week period. This does not account for peak concentrations which could precipitate respiratory symptoms. If the mean and maximum value is compared for the three main compounds with published safety levels, a slightly different picture emerges. The 2003 WHO guideline level for Benzene is a running annual mean of $16.25\,\mu g\,m^{-3}$ reducing to $3.25\,\mu g\,m^{-3}$ by 2010. The maximum mean concentration over a four-week period was measured at $93.5\,\mu g\,m^{-3}$ in one dwelling. If this level was maintained for a 12-month period, it equates to between 6 and 30 times the level of exposure recommended by the guideline. As this was the mean for a four-week period it will, at best, underestimate short-term peak concentrations. Similarly with toluene, which has a safety level of $260\,\mu g\,m^{-3}$ for one week, the survey found concentrations of up to $1783\,\mu g\,m^{-3}$ for a four-week period; approximately double the guideline exposure level, which itself is currently under review as the bio-science develops. M/p-Xylene has a guideline figure of $870\,\mu g\,m^{-3}$ for one year. The peak level found was $153\,\mu g\,m^{-3}$ for a four-week period. If a multiplier is used, to account for the equivalent of a 12-month exposure time span, a figure of $c.\ 1836\,\mu g\,m^{-3}$ is derived; again more than double the guideline. This high level may only occur in a relatively small number of dwellings, but no attempt was made to measure the short-term peaks which may be the most influential factor in triggering respiratory symptoms in atopic individuals. Calculating the geometric mean using absorbent tube technology is a rather blunt instrument if clinical symptoms are important primary outcomes.

Longitudinal monitoring on a much shorter time frame is required to check for incidents that could trigger asthmatic attacks. Molhave and Thorsen[51] for instance, observed a 16-fold increase in VOC levels with the re-activation of the HVAC system after night shutdown. They postulated that the dust in the ducts serves as a VOC sink. Such an increase could provoke symptoms in asthmatics or those with multiple chemical sensitivity.

Pesticides

Pesticides are toxic substances deliberately introduced into the domestic environment to control pests such as moths, fleas, cockroaches, flies, ants, rodents and even moulds and bacteria. Godish[52] has identified the most commonly used indoor pesticides as hypochlorites, ethanol, isopropanol, pine oil, glycolic acid, 2-phenylphenol and 2-benzyl-4-chlorophenol. There is also a variety of volatile organic insecticides which are used to treat building materials. The most common are: diazinon, chlordane, lindane, malathion, chlorpyrifos, aldrin and boric acid. These are introduced either as emulsion sprays or fogging devices (mothballs), impregnated into pest or bait strips or incorporated in flea/tick collars. Those pesticides with high vapour pressures (c. 1.2×10^{-2} mm Hg), off-gas more easily and can be found in greater concentrations, particularly immediately after application.

Implications for occupant health are as short-term respiratory tract or skin irritants and long-term carcinogens. Compounds such as chlordane, heptachlor, lindane, dieldrin and PCP are known to accumulate in fatty tissue and are suspected human carcinogens. The widespread use of 'Halophane' washes associated with inhibiting and/or treating mould growth in a large number of Scottish dwellings suffering from winter condensation, undoubtedly constitutes a significant exposure hazard. Acaricides which are sometimes used to control HDM populations, are toxic pesticides. They are used with the optimistic intention that they prove more toxic to the HDM than to the dwelling's human occupants. In general the extent of pesticide contamination and potential health risks can be reduced by minimising their use and providing high ventilation rates immediately after application, when concentrations will peak.

Toxic chemical particulates

Godish[52] has maintained that there is increasing epidemiological evidence that particulate phase matter (typically referred to as 'dust') is a major risk factor in occupant health. The generic term 'dust' refers to a large range of particle sizes, types, sources and chemical compositions. It can be measured using a number of techniques which count total suspended particles (TSP), respirable suspended particles (RSP), UV particulate matter (UVPM), particle numbers or specific particle fractions. Particles less than 3 microns are the most biologically significant as they can penetrate much deeper into the lungs.

Particulates from engine exhausts which can be found in large concentrations near main roads (particularly on city centre uphill bus routes) are of a size that can accumulate in the lung, as the macrophages are not sensitive to such tiny diameters. There is thus a concern that increasing ventilation rates in areas where external ambient pollution is high, due either to high traffic densities and low traffic speed, or where smoke stacks from incinerators regularly ground in certain weather conditions, will trigger respiratory symptoms. Van Steenis[53] for instance, has claimed that the switch to burning residual oils, increases the weight of nickel emissions by a factor of 55 and nickel, platinum, vanadium, mercury, lead and cobalt are also significant in the development and incidence of respiratory disease; as is insoluble uranium oxides, beryllium, quartz dust and acidic carbon. He claims that there are clear correlations between smoke stacks from oil refineries and incinerator chimney flumes – grounding at an average distance of 17 miles – leading to acute incidents of respiratory disease, heart attack and cancer. It would be prudent to switch-off any mechanical air intakes during such events. In general, indoor air pollution can normally be attributed to interior emission sources where the exposure duration is more prolonged.

Endotoxins

Among the indoor pollutants that may be implicated in the aetiology of asthma are bacterial endotoxins. These are lipopolysaccharides that are found in the cell wall of Gram-negative bacteria. In contrast to other indoor-related agents the methodology for measuring endotoxin exposure levels is not well defined. Simpson et al.[54] claimed that they have pro-inflammatory effects and have causative associations with a range of occupational lung diseases. In the homes of adult asthmatics Michel et al.[55] found a clear correlation between higher levels of indoor endotoxins and disease severity. Park et al.[56] suggested that exposure to relatively low levels of endotoxins in the first year of life may cause airway inflammation and wheeze in some infants. Reidler et al.[57] on the other hand, in a study of children living on farms, suggested a protective effect from high exposure levels when young. Platts-Mills et al.[58] have claimed that exposure to cats in childhood also appears to have a similar protective effect with less likelihood of atopic disease. He postulated a biphasic effect – early exposure being in some way protective, while prolonged exposure in later life, being pro-inflammatory. Indoor endotoxin levels appear to be strongly associated with pet-keeping.

A study by Tavernier et al.[59] in Manchester, UK, found high endotoxin levels in living room carpets at tenfold higher concentrations than bedroom carpets or mattresses (living room mean of 7 413 ng (spread 2798–19 633 ng): bedroom mean 428 (spread 217–843): mattress mean 198 ng (spread 105–361 ng). The mean value for asthmatic cases was almost double that of the healthy controls. The higher levels found in living room carpets were put down to heavier contamination by footwear, usage, food and drink spillage and pet occupation. The study hesitated to make any claims as to causality, as endotoxin levels may simply be correlated with some other environmental factor which was not measured. There is however, an implication that floor surface coverings and soft furnishings should be made easier to clean, which will inhibit endotoxin deposition and growth.

Dwellings that are damp or have signs of dampness related to a water incursion, appear to provide a range of environmental nutrients for microbial growth. A study by Huang et al.[60] found geometric mean endotoxin levels to be three times higher in dwellings with a history of water damage ($77.32–25.24 \, \mathrm{ng \, mg^{-1}}$ of dust). Although the sample was relatively small (45 dwellings) there were no other characteristics that appeared to be significant e.g. building type and age, use of air conditioner/dehumidifier/ air filter and even pet ownership did not appear to elevate endotoxin levels. Milton[61] reviewed 115 papers dealing with indoor endotoxins, and concluded that inhaling endotoxins at the upper end of normal domestic exposure levels, produces systemic symptoms and an increase in asthma severity. The USA Institute of Medicine's[62] conclusions on endotoxins were equally erudite:

> "Given the significant body of data on the exquisite sensitivity of the innate immune system to small quantities of endotoxin, the hypothesis that domestic endotoxin exposure may influence the development of the immature immune system or affect the severity of asthma, warrants further investigation."

Developing guidelines for domestic indoor air quality

In contrast to the domestic sector there are both guidelines and regulations issued annually by the Health and Safety Executive[63] for air quality in the workplace, which are based on the concept of occupational exposure limits (OEL), maximum exposure limits (MEL) and time weighted averages (TWA). These lists have legal status but are not exclusive. Absence from the list does not imply that a substance has no ill effects on health, nor that it is safe to use without control. Where a substance does not have an OEL the employer must carry out a risk assessment to determine an 'in-house' OEL.

The WHO has published guidelines[64] for air quality which are targeted at ambient air pollutants. They are based on inhaled exposure to a single airborne chemical. They do not take account of additive, synergistic or antagonistic effects, or exposure through routes other than the lungs. The guidelines provide information on sources, concentrations, proven and suspected health effects and health risks.

The UK Building Regulations are geared to prescriptive measures to ensure 'adequate ventilation' on the basis that the outdoor air is unpolluted, but there are currently no standards covering the quality of indoor air with respect to specific toxic pollutants known to derive from indoor sources. Harrison[65] has reviewed the guideline values now emerging in Canada, Finland, Germany and Norway. In general these are based on exposure levels (ppm or $\mu g\,m^{-3}$) over a given time span (5/30 min, 1–8 hr and annual means). In 1998, a Department of Health committee (UK), published a report[66] on the medical effects of air pollution on health. It concluded that there was no known safe threshold value for PM_{10} particulates, with each rise of $10\,\mu g\,m^{-3}$ responsible for a 3% increase in asthmatic attacks, bronchodilator use and lower respiratory tract problems and a 1.9% increase in acute hospital admissions.

Why set indoor air quality guidelines?

Harrison[65] gives five main reasons why guidelines need to be developed:

1   To inform the design and management of buildings.
2   To elicit action to reduce exposure to potentially harmful substances.
3   To indicate the need, if any, for the development of control or mitigation policies or regulations.
4   To facilitate the setting of appliance and/or product emission standards to help control pollution at source.
5   To underpin the development of information and advice.

The question of what pollutants should be monitored remains problematic. Of the studies which have attempted a multi-factorial approach, they have been unable to differentiate the influence, or even produce a rank order of the variables, likely to be the most significant in the aetiology or exacerbation of respiratory disease. It is clearly difficult – if not prohibitively expensive – to attempt to monitor everything. The best approach may be to develop a large enough evidence base where algorithms or stochastic modelling can be developed linking key known indoor pollutants with easily identified markers such as CO_2, NO_2, formaldehyde, allergens and moulds. Measuring temperature and humidity is also likely to be useful, not just in predicting the incidence of HDM allergens, mould and dampness, but the rate of VOC off-gassing from indoor sources. Appropriate and standardised measurement techniques will also have to be established, as will health risks from pollutants both in isolation and in combination. The state has legislated for health and safety in the workplace and in the car. Why not the home?

Summary

This chapter has considered some of the evidence base now emerging that is pertinent to indoor air quality issues. It is becoming clear that indoor air can contain a complex array of both naturally occurring and synthetic compounds, particulates and gases, as well as microbes and endotoxins. Reducing ventilation rates is more than likely to result in an increase in indoor pollutants, whether they are off-gassing from building components, fittings or fixtures; being generated by biological sources, or are as a result of occupant behaviour. The trend to make dwellings 'tighter', driven by energy-efficiency concerns, is again implicated in the production of a more toxic and unhealthy indoor environment.

References

[1] Raw GJ. Sick Building Syndrome: A review of the evidence on causes and solutions. HSE contract research report no. 42, British Research Establishment (BRE), Garston, Watford, 1992.

[2] World Wide Fund for Nature – Enough to make the blood boil, Initial result summary reported in the Scotsman, Science and Technology section, Edinburgh, 14th November 2003, p. 13.

[3] World Health Organisation, Indoor Air Pollutants, Exposure and Health Effects Assessment, Euro-reports and Studies no. 78, Copenhagen, 1983.

[4] Morrow LA, Ryan CM, Goldstein A and Hodgson MJ. A distinct pattern of personality disorder following exposure to mixtures of organic solvents, *Journal of Occupational Medicine*, Vol. 31, 1989, pp. 743–746.

[5] Ashford NA and Miller CS. *Chemical Exposures: Low Levels and High Stakes*, van Nostrand Reinhold, New York, 1991.

[6] Bell IR, Miller CS and Schwartz GE: An olfactory-limbic model of multiple chemical sensitivity syndrome: Possible relationships to kindling and affective spectrum disorders, *Biological Psychology*, Vol. 32, 1992, pp. 218–242.

[7] Carter MC, Perzanowski MS and Raymound A. Home intervention in the treatment of asthma among inner-city children, *Journal of Allergy and Clinical Immunology*, Vol. 108, No. 5, pp. 743–737.

[8] Bascom R, Kesavanathon J and Swift DL. Indoor Air pollution: Understanding the Mechanisms of the Effects, in RB Gammage and AB Berven (eds) *Indoor Air and Human Health*, ISBN 1–56670–144–9, 1996.

[9] Bascom R. Air pollution, in N Mygind and RM Naclerio (eds). *Allergic and Non-allergic Rhinitis*, Clinical Rhinitis, Copenhagen, Munksgaard, 1993, pp. 32–45.

[10] ASHRAE. Standard 62-1992, Ventilation for Acceptable Indoor Air Quality, American Society of Heating, Refrigeration, and Air Conditioning Engineers Inc, 1992.

[11] Carrer P and Maroni M. Allergens in indoor air: environmental assessment and health effects. *Science of the Total Environment*, 2002, Vol. 270, pp. 1–3.

[12] Kim CS, Lim YW, Yang JY, Hong CS and Shin DC. Effects of indoor CO_2 concentrations on wheezing attacks in children. The International Academy of Indoor Air Sciences, Monterey. *Indoor Air*, July 2002, Vol. I, pp. 492–497.

[13] Corsi RL, Torres VM, Sanders M and Kinney KA. Carbon dioxide levels and dynamics in elementary schools: Results of the TESAIS Study, Conference Proceedings. The International Academy of Indoor Air Sciences, Monterey, *Indoor Air*, July 2002, Vol. II, pp. 74–79.

[14] Sowa J. Air Quality and Ventilation Rates in Schools in Poland: Requirements, Reality and Possible Improvements. The International Academy of Indoor Air Sciences, Monterey. *Indoor Air*, July 2002, Vol. II, pp. 68–73.

[15] Wang Li JI and Zhao CY. A case of sick building syndrome caused by incorrect ventilation design of the tight building, Conference Proceedings. The International Academy of Indoor Air Sciences, Monterey. *Indoor Air*, July 2002, Vol. IV, pp. 490–493.

[16] Health Canada, Indoor Air Quality in Office Buildings: A Technical Guide. Ministry of Supply and Services, Canada, 1995.

[17] World Health Organisation, Guidelines for Air Quality, WHO, Geneva, 2000.

[18] Raw GJ, Coward SKD, Llewellyn JW, Brown VM, Crump DR and Ross DI. Indoor Proceedings, Indoor Air 2002, The International Academy of Indoor Air Sciences, Monterey, July 2002, Vol. IV, pp. 461–466.

[19] Morgan WJ and Martinez FD. Risk factors for developing wheezing and asthma in childhood. *Pediatric Clinician of North America*, Vol. 39, No. 6, 1992, pp. 1185–1203.

[20] Vork KL, Broadwin RL and Lipsett MJ. Household Environmental Tobacco Smoke (ETS) – Exposure and risk of childhood asthma – Techniques to reduce between study hetrogeneity in a meta-analysis. The International Academy of Indoor Air Sciences, Monterey. *Indoor Air*, July 2002, Vol. II, pp. 483–488.

[21] OEHHA. All chronic reference exposure levels adopted by OEHHA as of December 2001. Office of Environmental Health Hazard Assessment, California Environmental Protection Agency, 2001.

[22] Smith KR. Fuel combustion, air pollution exposure and health: The situation in the developing countries. *Annual Review of Energy and the Environment*, 1993, Vol. 18, pp. 529–566.

[23] Lai ACK, Thatcher TL and Nazaroff WW. Inhalation transfer factors for air pollution health risk assessment. *Journal of Air and Waste Management Association*, 1996, Vol. 50, pp. 1688–1699.

[24] Layton DW. Metabolically consistent breathing rates for use in dose assessments. *Health Physics*, 1993, Vol. 64, pp. 23–36.

[25] Nazaroff WW and Singer BC. Inhalation of hazardous air pollutants from environmental tobacco smoke in US residences. The International Academy of Indoor Air Sciences, Monterey. *Indoor Air*, July 2002, Vol. II, pp. 477–482.

[26] CIBSE Guide Section A2, Chartered Institution of Building Service Engineers, London, 1986, Table 1.10.

[27] Andrae S, Axelson O, Bjorksten B, Fredriksson M and Kjellman N. Symptoms of bronchial hyper-reactivity and asthma in relation to environmental factors. *Archives of Disease in Childhood*, 1988, Vol. 64, pp. 473–478.

[28] Dekker C, Dales R, Bartlett S, Brunekeef B and Zwanenburg H. Childhood asthma and the indoor environment. *Chest*, 1991, Vol. 100, pp. 922–926.

[29] Jaakkola J, Jaakkola N and Ruotsalainen R. Home dampness and moulds as determinants of respiratory symptoms and asthma in pre-school children. *Journal of Exposure Analyses of Environmental Epidemiology*, 1993, Vol. 3, pp. 129–142.

[30] Maier WC, Arrighi HM and Morray B, Llewellyn C and Redding GJ. Indoor risk factors for asthma and wheezing among Seattle school children. *Environmental Health Perspectives*, 1997, Vol. 105, pp. 208–214.

[31] Macher JM. Inquiries received by the Californian Indoor Air Quality programme on Biological contaminants in buildings. *EXS Advances in Aerobiology*, 1987, Vol. 45, pp. 275–279.

[32] Croft WA, Jarvis BB and Yatawara CS. Airborne outbreak of trichothecene toxicosis. *Atmospheric Environment*, 1986, Vol. 20, pp. 549–552.

[33] Mason CD, Rand TG and Oulton M, MacDonald J and Anthes M. Effects of *Stachybotrys chartarum* on surfactant convertase activity in juvenile mice. *Toxicology and Applied Pharmacology*, 2001, Vol. 172, pp. 21–28.

[34] Hyvarinen A, Pekkanen J, Halla-aho J, Husman T, Korppi M and Nevalainen A. Moisture damage at home and childhood asthma. The International Academy of Indoor Air Sciences, Monterey. *Indoor Air,* July 2002, Vol. I, pp. 467–471.

[35] Huttunen K, Hyvarinen A, Nevalainen A, Komulainen H and Hirovonen MR. Indoor air bacteria induce more intense production of inflammtory mediators than fungal spores in mouse and human macrophages. The International Academy of Indoor Air Sciences, Monterey. *Indoor Air,* July 2002, Vol. III, pp. 67–70.

[36] Fung F and Hughson WG. Health effects of indoor fungal bioaerosol exposure, Indoor Air 2002, The International Academy of Indoor Air Sciences, Monterey, July 2002, Vol. III, pp. 46–51.

[37] Brundage JF, Scott RM, Lednar WM, Smith DW and Miller RN. Building associated risk of febrile acute respiratory diseases in army trainees, *JAMA,* 1988, Vol. 259, pp. 2108–2112.

[38] Mendell MJ, Fisk WJ, Kreiss K, Levin H, Alexander D, Cain WS, Gimam JR, Hines CJ, Jensen PA, Milton DK, Rexroat LP and Wallingford KM. Improving the health of workers in indoor environments: Priority research needs for a national occupational research agenda. *American Journal of Public Health,* September 2002, Vol. 92, No. 9, pp. 1430–1440.

[39] Peden DB, Setzer RW, Jr and Devlin RB. Ozone exposure has both a priming effect on allergen induced responses as well as an intrinsic inflammatory action in the nasal airways of perennially allergic asthmatics. *American Journal of Respiratory Critical Care Medicine,* 1995, Vol. 151, pp. 1336–1345.

[40] McConnell R, Berhane K, Gilliland F, London SJ, Islam T and Gauderman WJ. Asthma in exercising children exposed to ozone: A cohort study. *The Lancet,* 2002, Vol. 359, pp. 386–391.

[41] Monn C. Exposure assessment of air pollutants: A review on spatial hetrogeneity and indoor/outdoor/personal exposure to suspended particulate matter, nitrogen dioxide and ozone. *Atmospheric Environment,* 2001, Vol. 35, pp. 1–32.

[42] Devlin RB, McDonnell WF, Mann R, Becker S, House DE, Schreinemachers D. and Koren HS. Exposure of humans to ambient levels of ozone for 6.6 hours causes cellulal and biochemical changes in the lung. *American Journal of Respiratory Cell Molecular Biology,* 1991, Vol. 4, pp. 72–81.

[43] Bayer CW, Cook AL, Roberts D, Via PD and Teague WG. Airs system: A real time monitor to measure indoor air pollutants and lung function in children with asthma. Conference Proceedings. The International Academy of Indoor Air Sciences, Monterey. *Indoor Air,* July 2002, pp. 498–502.

[44] Barry BE, Chang MP, Bloom SB, Moss NE and Bates CB. Assessment of exposures to airborne toxic chemicals and the development or exacerbation of asthma in the general population. Conference Proceedings. The International Academy of Indoor Air Sciences, Monterey. *Indoor Air,* July 2002, Vol. I, pp. 125–130.

[45] Brugge D, Vallaaarino J, Ascolillo L, Osgood ND, Steinbach S and Spengler J. Environmental factors for asthmatic children in public housing. The International Academy of Indoor Air Sciences, Monterey. *Indoor Air,* July 2002, Vol. II, pp. 428–431.

[46] Crump D. Minimising the impact on the indoor environment of chemical emissions from building products. International Academy of Indoor Air Sciences, Edinburgh. *Indoor Air,* 1999, Vol. 5, pp. 288–293.

[47] Coward SKD, Brown VM, Crump DR, Raw GJ and Llewellyn JW. Indoor air quality in homes in England, Volatile organic compounds, BRE, Watford, 2002.

[48] Jaakkola JJK, Spengler JD, Parise H, Kislitsin V and Lebedeva NV. Asthma, asthma-like symptoms and allergies in Russian school children in relation to new surface materials in the home. The International Academy of Indoor Air Sciences, Monterey. *Indoor Air*, July 2002, Vol. I, pp. 478–483.

[49] Jaakkola JJK, Oie L, Nafstad P *et al*. *American Journal of Public Health*, February 1999, Vol. 89, No. 2, p. 188.

[50] Rumchev K, Spickett J, Phillips M and Stick S. Conference Proceedings, Indoor Air 2002, The International Academy of Indoor Air Sciences, Monterey, July 2002, Vol. I, pp. 472–476.

[51] Molhave L. and Thorsen M. A model for investigations of ventilation systems as sources of volatile organic compounds in indoor climate. *Atmospheric Environment*, Vol. 25A, 1991, pp. 241–249.

[52] Godish T. *Sick Buildings: Definition, Diagnosis and Mitigation*, Lewis Publishers, 1995, ISBN 0–87371–346–X.

[53] Van Steenis D. Airborne Pollutants & Acute Health Effects, *The Lancet*, Vol. 345, April 1995.

[54] Simpson JCG, Niven RML, Pickering CAC, Fletcher AM, Oldham LA and Francis HM. Prevalence and predictors of work related symptoms in workers exposed to organic dusts. *Occupational Environmental Medicine*, 1998, Vol. 55, pp. 668–672.

[55] Michel O, Kips J, Duchateau J, Vertongen F, Robert L, Collet H, Pauwels R and Sergysels R. Severity of asthma is related to endotoxin in house dust. *American Journal of Respiratory Critical Care Medicine*, 1996, Vol. 154, pp. 117–120.

[56] Park JH, Gold Dr, Spiegelman DL, Burge HA and Milton DK. House dust endotoxin and wheeze in the first year of life. *American Journal of Respiratory and Critical Care Medicine*. Vol. 163, No. 2, 2001, pp. 322–328.

[57] Reidler J, Eder W, Oberfeld G and Schreuer M. Austrian children living on a farm have less hay fever, asthma and allergic sensitisation. *Clinical Experimental Allergy*, 2000, Vol. 30, pp. 194–200.

[58] Platts-Mills TAE, Custis N, Erwin EA, Sporik R and Woodfolk JA. Asthma and Indoor Air. Indoor Air 2002, The International Academy of Indoor Air Sciences, Monterey, July 2002, Vol. III, pp. 10–16.

[59] Tavernier G, Fletcher GD, Francis HC, Oldham LA, Fletcher AM, Stewart L, Gee I, Watson A, Frank TL, Frank P, Pickering CAC and Niven RML. Endotoxin exposure in asthmatic children and matched healthy controls: Results of IPEADAM study. Indoor Air 2002, The International Academy of Indoor Air Sciences, Monterey, July 2002, Vol. I, pp. 488–491.

[60] Huang YJ, Huang JY, Lin LL, Wu PC, Ma YP and Su HJ. Environmental endotoxin concentrations are associated with home characteristics. Indoor Air 2002, The International Academy of Indoor Air Sciences, Monterey, July 2002, Vol. IV, pp. 764–769.

[61] Milton DK. Bacterial Endotoxins: A review of health Effects and Potential Impact in the Indoor Environment, in RB Gammage and AB Berven (eds), *Indoor Air and Human Health*, Boca Raton, CRC Press, London, ISBN 1–56670–144–9, 1996.

[62] Institute of Medicine, Clearing the Air, Committee on the Assessment of Asthma and Indoor Air, National Academic Press, Washington, 2002, ISBN 0–309–06496–1.

[63] Health and Safety Executive, Occupational exposure limits EH40, HSE Books, 2003, PO Box 1999, Sudbury.

[64] WHO – Air Quality Guidelines for Europe (European Series No. 91), 2000, Copenhagen, Denmark.

[65] Harrison PTC. Guidelines for Indoor Air Quality in the Home, Unhealthy Housing: Promoting Good Health, Conference, Warwick University, March 2003, p. 8.

[66] COMEAP. The Quantification of the Effects of Air Pollutants on health in the United Kingdom, Department of Health, 1998.

Chapter 10

Air tightness and ventilation rates

One of the main confounding variables previously identified, is the influence and significance of background ventilation rates on the internal humidity, indoor air quality and the efficacy of any mechanical ventilation system. Although Chapter 5 attempted to model the possible scale of any historical reductions in ventilation rates during the twentieth century, these figures are at best indicative of a trend. Work undertaken worldwide has generated international comparisons[1-4] demonstrating a variation of up to $5\,ach^{-1}$ between background rates in the United States, Scandinavia and China. The blow door testing (described in Chapter 5) of five generic house types common to west central Scotland, produced a differential of the same order.

Air leakage rates in the UK

There are only three main databases on air leakage rates in UK dwellings. One is held by British Gas plc[5] and covers some 200 dwellings, the other two by the BRE[6,7] covering 471 dwellings and 87 large panel flats. Both reports used the 'blow door' methodology to estimate air leakage rates as a proxy for air infiltration rates. It must be recognised that due to the design of openings and components, air leakage rates measured when the dwelling is internally pressurised, may not necessarily mimic air infiltration rates. External wind pressure will be directional and induce both positive and negative zones around the dwelling's periphery. The BRE report recognises that the aim of improving air tightness is to retain heat, improve comfort and economy and reduce CO_2 emissions arising from space heating. The test programme, which pressurised dwellings to 50 Pa, confirmed the variability in air leakage rates with the most 'leaky' being over ten times that, of the most air 'tight'. The report also concluded that, 'it is currently impossible to make a realistic estimate of air tightness of a dwelling, newly built or otherwise, by simple inspection alone; some form of measurement being required'.

While this appears to be true of the stock for the greater part of the twentieth century, the report did highlight that dwellings built after 1987 were significantly more air-tight than their predecessors. Recent changes to the Building Standards requiring pressure testing to demonstrate 'tight construction' may become a powerful driver in changing construction techniques. The house builders have, in general, avoided pressure testing by use of a legal loophole created by the use of the word 'capable' in the regulations. As the Swedish standard[8] expresses the requirement as a maximum leakage of $3 \, m^3/m^{-2}/hr$, it is obvious that the existing British stock is nowhere near this standard. The mean air leakage rate at 50 Pa for the 471 dwellings tested[6] was $13.1 \, ach^{-1}$ with a range of 2–3 at the tight end and 29–30 at the leakiest. For dwellings constructed since 1987, the mean was 9.6, with a range of 3–4 and 21–22 ach^{-1} at 50 Pa.

Where flues and chimneys have been tested[9] a flow rate of $432 \, m^3/hr$ at 50 Pa was measured for an open fire of approximately 400 mm, with an 8-m chimney. In a room with a volume of less than $50 \, m^3$ this represents an air change rate of over $8 \, ach^{-1}$.

Air change rate and meso-climate

Although 50 Pa is an extreme pressure, wind speeds of up to gale force measuring eight on the Beaufort scale (equating to $17.2–20.7 \, ms^{-1}$) are not unusual in urban west central Scotland[10] where the average wind speed is between 6 and $8 \, ms^{-1}$ and wind speeds of between 9 and $10 \, ms^{-1}$ are not uncommon (10% frequency). Without complex dynamic simulation which would take into account: building volume-to-area ratio, aspect, prevailing wind, shelter, height, number of storeys, leakage to adjacent dwellings, construction techniques, detailing and opening characteristics, it is impossible to calculate the air change rate that will be induced by external wind pressures. From the ESP-r modelling in Chapter 5, changes in wind speed and direction had a significant influence on air change rates over the 48-hr period (Figures 10.1 and 10.2 for the 1890s tenemental model).

A change of wind speed from c. 3 to $7 \, ms^{-1}$ produced an increase in the internal air change rate of c. $1.2 \, ach^{-1}$ ($0.78–2 \, ach^{-1}$). When expressed as a crude ratio, it appears that an increase in wind speed of $1 \, ms^{-1}$ (accepting that direction will also be influential) equates to an increase of approximately $0.25 \, ach^{-1}$. A room with an open flue may thus have an additional two air changes induced, when the external wind speed increases from flat calm to $8 \, ms^{-1}$, and the occasional gale force conditions could more than double this figure, if the relationship remains linear. Comparing the 1890s tenement with the contemporary timber-frame model – which has lightweight construction and no chimney – the external wind conditions are less influential in driving internal air change rates, with a maximum differential of $0.4 \, ach^{-1}$ being induced by changes in wind speed and direction. Norm referencing these two simulations produces the somewhat unsurprising outcome that external weather conditions can affect properties with open chimneys

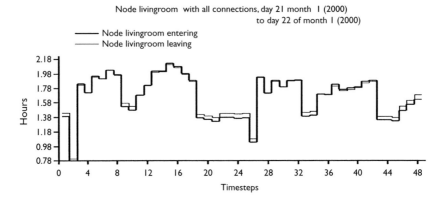

Figure 10.1 Effect of wind speed and direction on air change rates (1890s model).

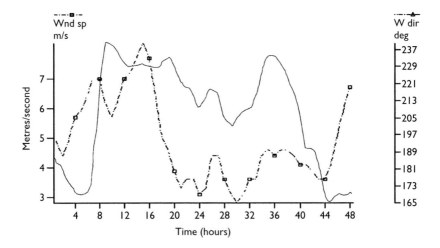

Figure 10.2 Wind speed and direction.

to a far greater degree (factor of three) than relatively 'tight' dwellings (Figures 10.3 and 10.4).

There is clearly a large number of variables that do not allow air leakage characteristics to be easily converted to air infiltration rates, and air speeds and pressure do not have a linear relationship (a doubling of air speed requires the pressure to be squared). Stephen[6] claims that there is a widely used rule of thumb whereby average natural ventilation rates in a given dwelling, represent one-twentieth of the air leakage rate at 50 Pa. Stephen concedes that unpublished modelling work by the BRE has undermined this ratio, and varying some of the above factors could change the divisor to between

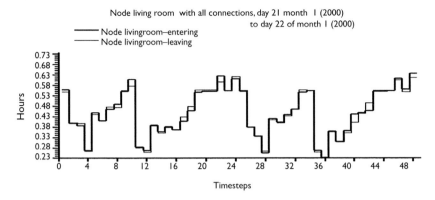

Figure 10.3 Effect of wind speed and direction on air change rate (2000s model).

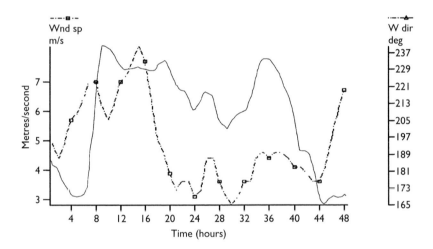

Figure 10.4 Wind speed and direction.

10 and 30. This represents a large margin of error and in turn will have a major influence on both internal comfort conditions and/or heating costs.

Ventilation rates and energy efficiency

The energy audits carried out on the stock involved in the interventionist trial reported in Chapters 6 and 7 produced a mean NHER of 5.2. This is better than the Scottish average of 4.1, generated by the 1996 Scottish House Condition Survey. The air change rates contained in the calculations estimate an average of 0.88 ach^{-1} with a range of 0.62–1.15 ach^{-1}. Such rates equate to *c*. 70 W/K (range 33–151 W/K). When compared with the average fabric losses of 173 W/K (range 63–274 W/K) this rate represents *c*. 40% of the specific heat loss. The calculations could thus have underestimated the overall heat loss – particularly during windy episodes – however, the dwellings did have some modifications that could render them more airtight than those previously tested.

Etheridge[5] found that cavity fill materials, (foam fill in particular), offer improvements in air tightness by sealing cracks in the building fabric. The type of wall construction had a major influence on air tightness when pressurised to 50 Pa, with large panel systems (LPS) the tightest at 7.5, timber frame at 8, solid masonry at 11.5 and cavity masonry was measured at just over 14 ach^{-1}. The LPS dwellings appeared to be more airtight as the heavy weight precast concrete panels themselves are relatively dense with leakage only occurring at joints, service entries and windows. This contrasts with cavity masonry construction where 71% of the total loss was down to a myriad of cracks and openings throughout the construction.

Of the 34 dwellings that completed the interventionist study, 22 had urea-formaldehyde cavity insulation foam and 21 had new double-glazed PVCu tilt and turn windows, many of which did not incorporate trickle ventilators. Almost all dwellings had wall-to-wall fitted carpets or timber laminate on the original tongued and grooved floorboards. These features may have rendered them relatively airtight, however, without blow door testing, no conclusions can be drawn and this area remains one of the main confounders in assessing the efficacy of the MHRV intervention.

Tests by the BRE in conjunction with Heatwise Glasgow in the late 1980s on 22 Scottish 1960s tenemental flats (see Chapter 4, Figure 4.5) produced a figure of 21.3 ach^{-1} at 50 Pa. This was close to double their English counterparts and draught-stripping only produced a reduction of *c*. 5%, with cracks at the floor/wall and window/wall providing significant air leakage pathways. This is in line with the findings of the larger study[6] where the effectiveness and condition of draught-stripping was variable, with some draught-stripped windows being as leaky as well-fitting windows without draught-strips. Overall the air leakage rate attributable to windows was reported to be around 16%.

In the BRE study[6] of 471 dwellings, suspended ground floors tended to be more leaky than solid floors. The floor covering was influential in reducing air leakage (e.g. wall-to-wall carpet and underlay reduced pathways as did tongued and grooved floorboards). The number of storeys in a dwelling was not a significant factor. Although the study failed to provide conclusive evidence, there was a suggestion that air leakage rates increased in the winter as lower temperatures produced shrinkage in components and structure, thus allowing joints and cracks to increase in dimension. This effect is of some concern as a greater air infiltration rate could result during the colder months. This would also lead to questions concerning the veracity and replicability of inter-seasonal pressure testing.

'Build tight – ventilate right'

In 1992 Perera and Parkins[11] put forward the concept of 'Build tight – ventilate right'. This was a proposition that dwellings should be designed and constructed to be as tight as practicable and incorporate a planned ventilation strategy. This approach reflects and addresses several current concerns regarding indoor air quality, energy use and environmental issues. It emphasises that a building cannot be too airtight, but it can be under-ventilated.

The approach built on previous work published in 1985, where the BRE[12] claimed that there was wide acceptance that a ventilation rate of $0.5\,ach^{-1}$ – supplemented by air extraction during cooking and bathing – was generally adequate to keep indoor pollutant concentrations below acceptable maxima, as well as avoiding condensation. There is relatively little evidence to support this hypothesis and two obvious factors undermine the development of regulations based on such a dubious premise. Their own work on air leakage rates[6] produced a mean of $13.1\,ach^{-1}$. Approximately 71% of this air leakage was due to cracks in the construction. Dividing this average air leakage rate by 10:30 (the rule of thumb suggested by Stephen[6]) produces air change rates between 0.44 and $1.3\,ach^{-1}$. The BRE's air quality survey into indoor VOC concentrations[13] highlighted concerns over many compounds and ignored short-term maxima, which could be implicated in triggering acute respiratory symptoms. If such toxicity is prevalent in relatively 'leaky' dwellings (i.e. above the $0.5\,ach^{-1}$ figure) any further reduction will have a detrimental effect on indoor air quality. Such a low prescriptive rate does not appear to supply the 8 l/s per person recommended by CIBSE[14] for odour control, when more than one person occupies the average living room volume (8 l/s equates to $29\,m^3/hr$ – average living room volume of $50\,m^3 \times 0.5\,ach^{-1}$).

Mechanical heat recovery ventilation

The move towards wet zone extraction, now enshrined in the last two Building Standard (Scotland) Regulation revisions, requires mechanical extraction in the bathroom and kitchen as a 'deemed to satisfy' option. Although such a strategy should ensure that some of the water vapour from cooking and bathing is expelled at source, it also puts an onus on the occupier to use the systems in a prudent and sensible manner. Extracting air will also extract heat. As heat may be viewed by many as an expensive commodity, there is a temptation to ignore the fan option, as it has clear cost/benefit implications as well as providing an intermittent acoustic nuisance. It is therefore important that the occupants understand the ventilation provision – be it natural or mechanical – and the role indoor air quality plays in ensuring good respiratory health.

The research trial went to some lengths to educate the occupants as to the function and operating modes of the MHRV units. Measured consumer satisfaction was high and there were no complaints regarding fan noise. Only one of the 120 units fitted was sabotaged. These units – having a built-in thin film plastic heat exchanger with a claimed efficiency of c. 70% (the actual efficiency is probably closer to 50%) – did not produce a noticeable chilling effect when the room was occupied and heated. Occupants were invited to experience the temperature of the incoming airstream to convince themselves that the unit would not result in the depression of internal air temperatures.

The three half-house units that were fitted had additional benefits, providing discrete wet zone extraction and room supply. This larger heat exchanger in combination with lower air speeds enhances the efficiency of the heat transfer – the negative points being the installation of surface-mounted ductwork and the additional capital cost (c. £1500 compared with £500 for two cartridge fan units).

A BRE[6] report published in 1998 claimed that where continuously running balanced mechanical ventilation systems are provided, natural infiltration through the air leakage paths continues independently and will thus – in the average dwelling – lead to over-ventilation and energy wastage. This view appears to ignore or discount the benefits of cross-flow heat recovery technology. Such an approach can allow air change rates to be increased significantly without incurring a proportional heat loss penalty. Indeed the overall air change rate can be increased to c. $1\,\mathrm{ach}^{-1}$ for the equivalent heat loss stemming from natural ventilation regimes taken to be $0.5\,\mathrm{ach}^{-1}$. In large or particularly leaky dwellings with open flues and ill-fitting windows, problems associated with under-ventilation and pollutant build-up are less likely.

If moisture extraction – or more accurately, RH suppression below 60% – becomes the primary driver of domestic ventilation, what should be the prescribed rate that will produce the optimum balance between energy efficiency and vapour control, for a given temperature? If an average whole-house temperature of 18°C is taken as a reasonable target, to maintain RH below 60%, the mixing ratio cannot rise above $8\,\mathrm{g\,kg}^{-1}$ of dry air. At 18°C, 1 kg of dry air has a volume of $1.21\,\mathrm{m}^3$.[14] The BRE[15] has calculated that the average daily moisture burden per household is between 7 and 14 litres of water vapour per day. For a dwelling with a volume of c. $200\,\mathrm{m}^3$ this equates to between 1.76 and 3.53 g of moisture/kg of dry air per hour. The ambient humidity will also contribute to the internal mixing ratio. To maintain the internal RH below 60% would thus require the ambient humidity to remain below $6.24\,\mathrm{g\,kg}^{-1}$ of dry air for the lower rate of moisture production (i.e. 7l per day) and $4.47\,\mathrm{g\,kg}^{-1}$ of dry air for the higher rate. These levels will reduce with increasing temperature.

As ambient absolute humidity will vary depending on latitude, longitude and meso-climatic factors, a dwelling's ventilation regime thus has to be designed for the locale, as well as the dwelling's volume and vapour production characteristics. It is thus difficult to produce a simple rule of thumb prescribing a universal air change rate. A more productive approach would be to produce a performance specification that requires internal RH to be maintained below the critical equilibrium threshold of 60%. Increasing internal temperatures will accommodate a greater moisture burden without an increase in RH. The air temperature, however, will have to be maintained over the entire 24-hr period, as any drop will lead to a proportional increase in the RH. Using mean temperature and humidity targets is also inappropriate as a large variation around the mean could result in the CEH threshold being breached for a significant period.

Domestic ventilation should thus be designed to avoid peak concentrations, implicated in triggering respiratory symptoms. To achieve this, it appears that the recommendation of $0.5\,\mathrm{ach^{-1}}$ will not be adequate.

The question remains as to what is the maximum ventilation rate that can be tolerated in terms of energy efficiency? It appears appropriate to adopt the 'precautionary principle'. Do not design down to a minimum; design up to a maximum. Discomfort from air movement being the main limiting factor. Such an approach may result in occasional over-ventilation in some 'leaky' dwellings, but the air quality in most small 'tight' dwellings will improve and such an approach will ensure that prolonged exposure to a range of hazardous airborne pollutants is less likely to occur.

Summary

The historical move towards small tight and warm dwellings – which appears to be gathering pace at the start of the twenty-first century – can produce an indoor environment that is likely to suffer from poor air quality along with high absolute humidity. In such circumstances the evidence is building that additional ventilation measures are required to prevent deleterious effects on occupants' health status from allergens, gases, toxins, microbes, bioaerosols, VOCs, ETS and PM emanating from many diverse sources.

The apparently ubiquitous nature of the HDM in domestic environments fits well with Stephen Jay Gould's[16] theory of punctuated equilibrium. A relatively rapid change has occurred in a specific habitat allowing one species to proliferate at a rate that has never before been possible. The human species is suddenly exposed to a specific allergen in concentrations which do not occur in the natural environment. When combined with the increase in atopy – possibly being driven by a wide range of factors (genetics, hygiene, high salt/fat diet, prophylactic use of antibiotics, reduction in breast feeding, immunisation, paracetamol etc.) – the immune system in many individuals has become hyper-sensitive, producing powerful allergic reactions to what is, in essence, a range of harmless substances.

The challenge is thus to create comfortable and energy-efficient living conditions, without producing the damp micro-climates where dust mite colonies can proliferate. A Cochrane review[17] of the randomised controlled trials on the use of humidity suppression measures for asthma control found that only one trial[18] could be included, and concluded that there was a need for further large-scale, double-blind, randomised controlled trials measuring clinical outcomes in patients with asthma. Such a trial went on-site in North and South Lanarkshire in October 2003 and will report in 2005. This study, similar to the first phase, will include strategies designed to reduce the existing reservoir of dust mite allergens (steam cleaning, encapsulation of mattresses in micro-weave envelopes, renewal of duvets and pillows) and an intervention designed to inhibit the growth of mite colonies by reducing internal water vapour pressures using a whole house MHRV system. The effect of the MHRV units on asthma control of patients allergic to HDM, will be examined by means of a randomised, triple-blind, placebo-controlled, interventionist trial designed to answer a range of research questions. This work is primarily concerned with remediating the existing stock, but what design strategies and techniques should be adopted for new build?

References

[1] Murray DM and Burmaster DE. Residential exchange rates in the United States: Empirical and estimated parametric distributions by season and climatic region. *Risk Analysis*, Vol. 15(4), pp. 459–465.

[2] Øie L, Stymne H and Boman C-A *et al*. The Ventilation Rate of 344 Oslo Residences, *Indoor Air*, 1999, Vol. 8, pp. 190–196.

[3] Stymne H, Boman C-A and Kronvall J. Measuring Ventilation Rates in the Swedish Housing Stock, Building and Environment, 1994, Vol. 29, No. 3, pp. 373–379.

[4] Tao R-G, Wang C-X and Geng J-X *et al*. Characteristics of the rural houses in Xuanwei, China in Lung cancer and indoor air pollution from coal burning, in He X-Z and Yang R-D (eds). Kunming, China: Yunnan Science and Technology Publishing House.

[5] Etheridge DW, Nevrala DJ and Stanway RJ. Ventilation in traditional and modern housing. Research and Development Division, British Gas plc, 1987.

[6] Stephen RK. Airtightness in UK dwellings: BRE's test results and their significance, BRE, Garston, 1998.

[7] Cornish JP, Henderson G, Uglow E, Stephen RK, Southern JR and Sanders CH. Improving the habitability of large panel system dwellings. BRE Report BR154. BRE, Garston, 1989.

[8] BFS. Airtightness and heat recovery. National Board of Housing and Planning, Karlskrona, Sweden, 1988, Vol. 18, Chapter 3, pp. 13–14.

[9] Basset M. Building site measurements for predicting air infiltration rates. ASTM Symposium on measured air leakage performance of buildings, Philadelphia, USA, 2–3 April, 1984.

[10] CIBSE A2-3. Weather and solar data, CIBSE, London, 1986.

[11] Perera E and Parkins L. Build tight – ventilate right, Building Services, June 1992, Chartered Institution of Building Services Engineers, London, pp. 37–38.

[12] BRE. Surface condensation and mould growth in traditionally built dwellings. BRE Digest 297, Garston, 1985.

[13] Coward SKD, Brown VM, Crump DR, Raw GJ and Llewellyn JW. Indoor air quality in homes in England, Volatile organic compounds, BRE, Watford, 2002.

[14] CIBSE Guide. Volume A, Design data, The Chartered Institution of Building Services Engineers, 5th Edition, London, 1974.

[15] Garratt J and Nowak F. Tackling condensation, Building Research Establishment, Garston, Watford, 1991.

[16] Gould SJ. Ever Since Darwin: reflections in natural history, Burnett books, 1978.

[17] Singh M, Bara A and Gibson P. Humidity control for chronic asthma (Cochrane Review). In: The Cochrane Library, Issue 2, 2002. Oxford: Update Software.

[18] Warner JA, Fredrick JM, Bryant TN, Weisch C, Raw GJ, Hunter C, Stephen FR, McIntyre DA and Warner JO. Mechanical ventilation and high frequency vacuum cleaning: A combined strategy of mite and mite allergen reduction in the control of mite sensitive asthma. *Journal of Allergy and Clinical Immunology*, 2000, Vol. 105, pp. 75–82.

Chapter 11

Developing a new low allergen prototype dwelling

In the main, the challenge is to use this work and its primary focus on occupant health, to drive changes in domestic new-build design specifications and strategies. Such an undertaking has to address issues of indoor air quality (toxins and particulates), humidity suppression and the categorical imperative to design for a sustainable future, as it would clearly be iniquitous to improve the health of one population cohort at the expense of another. For this to happen, architects require to move away from their unfathomable predilection with vacuous modernist ideology and adopt the four 'Rs' of so-called 'green practice': reduce, reuse, recycle and reverse; reduce the demand for resources and the associated pollution; reuse existing materials and components where possible; design new components using recycled material in turn to be recyclable (the cradle-to-cradle scenario) and/or use materials that will eventually reverse to their original state, to reduce or ameliorate the collateral damage to the planet's ecosystems and biosphere. Failure to do so will justifiably see the profession of architecture further marginalised and impoverished.

The bio-climatic 'zoned' dwelling for northern latitudes

The design 'problem' is thus to develop proposals that produce the optimum achievable balance between energy efficiency and indoor air quality. Increasing ventilation rates to suppress RH below the 60% threshold required for HDM colonisation may increase energy usage and its associated deleterious pollution. The primary design principle is based on the concept of the 'zoned' dwelling; expanding the living area and volume when the climate allows and retreating into a well-insulated and buffered core during the coldest periods. To do this, the principles of timber frame construction have to be reversed. It is the 'tea cosy' strategy, a heavyweight thermal mass is externally insulated, to optimise the storage of solar and casual gains, using air transfer as the primary energy vector.

Thermal capacitance

Many involved with the design of the built environment do not appear to understand the concept of whole life-cycle analysis and erroneously an overemphasis on reducing the embodied energy associated with material selection and specification. This appears to be a laudable aim, however, as most of the energy/pollution burden over the building's lifetime will be energy in use (normally at a ratio greater than 25:1 – for Georgian/Victorian dwellings that have been around for 200 years the ratio could have climbed to greater than 200:1), increasing the embodied energy with the goal of minimising total life-cycle emissions will, in most cases, prove to be a more beneficial strategy.

Buildings constructed before the advent of the internal combustion engine had little or no associated CO_2 burden, as horsepower is CO_2 neutral. The transportation of heavyweight materials, although incurring associated hydrocarbon burdens, can provide a dwelling with vital thermal capacitance/inertia. As diurnal temperature fluctuations will normally result in a sharp increase in RH – particularly where the construction is lightweight timber frame – encapsulating a heavyweight core with a thick 'jacket' of insulation can store heat emanating from a variety of sources and reduce problematic internal temperature fluctuations. Such thermal inertia allows the energy flow to reverse from heavyweight mass to air, maintaining dry bulb temperatures and in turn ensuring RH does not increase overnight above the 60% CEH threshold for HDM activity. Additionally, such a strategy dispenses with the need for polythene vapour barriers as dew point will not normally be reached within the structural mass, and vapour diffusivity will progressively increase from the inner to outer layers, allowing any moisture to dissipate to the outside air.

Increasing the surface area of this thermal mass will further allow for efficient energy transfer of solar and casual gains. Pre-cast structural hypocaust floor slabs and 'rough' cavities are two techniques that can enhance efficient energy transfer into the core of the thermal mass. Heavy mass can also store 'coolth' in the summer, and when combined with MHRV systems can use diurnal heat flushing (circulating cool night air through the building fabric) as passive cooling during the July and August peaks. This warm thermal mass also maintains internal radiant temperatures encouraging the body's vasodilation system to expand the veins which in turn can lower blood pressure. As most of the energy is held in the thermal mass, occasional ventilation 'flushing' to extract smoke or cooking smells will not depress internal temperatures.

Insulation

As the overwhelming priority is to reduce CO_2 emissions from the dwelling over its expected lifespan, Eurisol[1] has claimed that the optimum 'environmental' thickness of insulation (comparing the embodied CO_2 associated

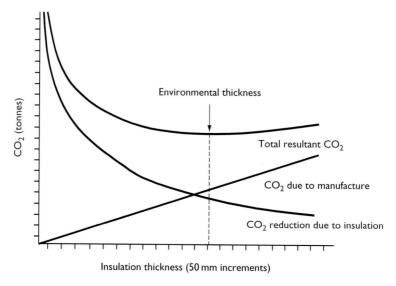

Figure 11.1 Embodied CO_2 vs CO_2 savings over 30-year lifespan (mineral wool).

with the production process with CO_2 savings in use, over a 30-year lifespan) should be between 800 and 900 mm (Figure 11.1).

Such a thickness, however, would provide significant design challenges and be somewhat profligate with land. From Figure 11.1 it is also clear that, as the saving is not linear, a thickness of 300 mm will provide a significant proportion of the energy savings (80%) with only relatively simple constructional modifications at sills and lintels being required. The use of semi-rigid wood-fibre-based products (such as Pavatex supplied by Natural Building Technologies[2]) slotted between an external timber 'web column/rafter' (Figure 11.2) allows for the use of local, low-quality softwood with no CO_2 penalty. The prefabricated system facilitates 'buildability', 'demountability' and eventually total reversibility, as most of the construction materials will biodegrade or break down through erosion.

For all glazed openings rigid insulation (Foamglas or similar) laminated between timber facings, can be used to back-insulate, significantly reducing the U-value of these elements (typically from c. 3.3 to 0.5 W m^{-2}/°C). Such a strategy can allow the energy balance through the element to be positive, capturing insolation during the hours of light and back insulating at night (i.e. a negative U-value when averaged out across the year – Figure 11.3).

Passive and active solar strategies

Direct gain sunspace: A sun-buffer zone is incorporated along the entire south elevation of the dwelling. Porteous[3,4] has demonstrated that such an element – if appropriately dimensioned and detailed – can typically displace

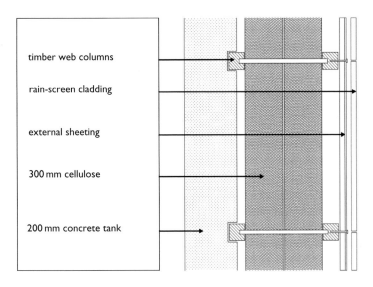

timber web columns	
rain-screen cladding	
external sheeting	
300 mm cellulose	
200 mm concrete tank	

Figure 11.2 Web beam/column detail.

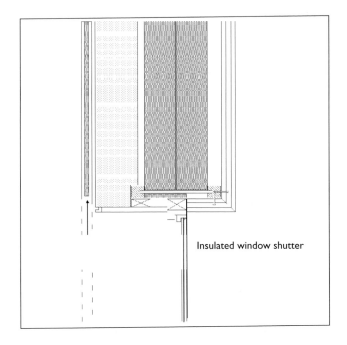

Insulated window shutter

Figure 11.3 Window back-insulated with sliding shutter.

c. 30–40% of energy demand in northern latitudes, by buffering the dwelling from diurnal back-losses and capturing and storing passive solar energy gains (Figure 11.4). The strategy is to maximise useful solar gains in the spring and autumn, thereby reducing the heating season to less than two or three winter months.

Air pre-heat for mass storage: A thermostat-controlled fan extracts warm air from the apex in the sun-buffer zone and delivers it to a 'hypocaust' floor or 'rough' cavity (incorporated in the north wall). The large surface area and thermal capacitance allows efficient low temperature energy to be transferred into the core of the thermal mass. One option would be to provide a controlled exhaust on the north elevation. Although a 'closed' loop system would normally be considered more efficient, any pollutants or dust build-up associated with ductwork – and implicated in SBS[5] – would be avoided, as all air is eventually expelled. The downside is that moisture from clothes drying in the sun-buffer zone could be periodically delivered into the floor. Passing the air through a heat exchanger incorporating filters would reduce the initial energy capture, but allow a closed loop system (Figure 11.5). Maintenance then becomes an issue and where possible, the air supply ducts should be short and easy to clean. The system can also be reversed in peak summer to store night coolth (Figure 11.6). This will become important with rising summer temperatures, if the CO_2 burdens associated with air conditioning are to be avoided. Current climate models are predicting summer design temperatures for south-east England of 38 °C

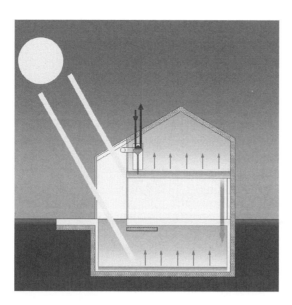

Figure 11.4 Direct solar gain to hypocaust storage.

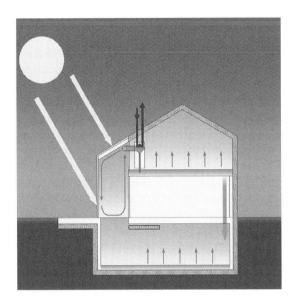

Figure 11.5 Indirect solar gain through MHRV unit to hypocaust storage.

Figure 11.6 Night cooling and through MHRV unit to hypocaust storage.

by 2050. California has recently experienced such a scenario with the generating and supply grid, unable to meet peak summer demand.

Domestic hot water (HW) solar pre-heat: Incorporating *c.* 6 m^2 of flat-plate solar collectors to shade the kitchen window will displace electricity or seasoned wood-pellet usage in the provision of domestic hot water. To maximise available solar flux, a pre-heat tank should be incorporated with a 'dublo' water vessel with thermostatic control valves separating the calorifiers. This will allow each tank to be heated in series, maximising hot water energy storage in the summer months while providing useful winter pre-heat. The integration of a multi-fuel top-up stove will compensate for the predictable reduction in mid-winter insolation. To provide a further 'fail-safe' option a simple electric 'booster' can make up any spring/autumn temperature deficit (Figure 11.7 for schematic). The system allows an 'energy manager' to draw from the warmest tank at any given time. In winter when the multi-fuel stove is operating, the solar collector simply acts as a pre-heat facility. The storage capacity can be geared to the likely demand as well as accounting for climate and orientation.

Photovoltaic niche application to sun-buffer-inclined glazing: As any useful increases in the sun-buffer air temperature will be driven by solar gains, incorporating transparent PV panels would support the use of DC fan power to transport warm air through the hypocaust, while shading the kitchen from direct gains. The system can thus be designed to be self-regulating, with air movement occurring and increasing in direct proportion to available solar flux. A limited battery storage capacity would also capture peak gains in high summer and could be used to drive nocturnal 'coolth' storage regimes or auxiliary DC lighting circuits. The equivalent of *c.* 6 m^2 of mono-crystaline panels could provide (at 15% efficiency) about 8100 W peak. At 240 V such an output could move over 1000 m^3 of air per hour (i.e. >2 ach^{-1}). With average solar flux at 100 W m^{-2} for the region, 90 W would be available during more overcast days. DC fan power is significantly more energy efficient and has the benefits of simple self-regulating controls.

Ventilation

The use of a whole-house MHRV system ensures vitiated and humid air extraction from the kitchen and bathroom while supplying 'fresh' pre-warmed air to the bedrooms and living room. This creates a constant negative air pressure in the 'wet zones' which should incorporate clothes-drying 'airing' cupboards. Alternatively, clothes drying can be undertaken in the sun-buffer zone when conditions support the required rate of evaporation. Such a system will also transfer energy from the multi-fuel stove and casual gains throughout the dwelling, with only 20–25% energy loss due to heat exchanger efficiency. The constant extraction of humid air at a controlled rate will suppress the internal mixing ratio that can then be maintained at close to ambient – between 3.5 and 7 g kg^{-1} of dry air for all but the peak summer

System design

Although hot water storage is treble the normal domestic minimum requirements, the system has been designed to produce fast and economic demand response across all seasons. All three cylinders can operate simultaneously at different working temperatures.

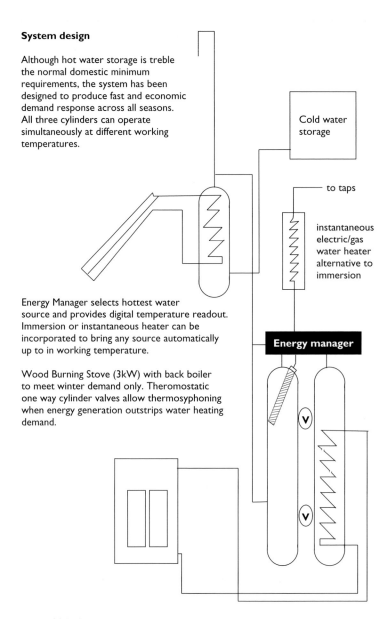

Cold water storage

to taps

instantaneous electric/gas water heater alternative to immersion

Energy Manager selects hottest water source and provides digital temperature readout. Immersion or instantaneous heater can be incorporated to bring any source automatically up to in working temperature.

Wood Burning Stove (3kW) with back boiler to meet winter demand only. Theromostatic one way cylinder valves allow thermosyphoning when energy generation outstrips water heating demand.

Energy manager

Figure 11.7 Solar hot-water pre-heat schematic.

months. Even allowing the internal whole-house temperature to drop to 18 °C will maintain RH close to 60% at a mixing ratio of $8\,g\,kg^{-1}$ of dry air. The rate of air extraction and input can be balanced and refined depending on seasonal and/or moisture/heat inputs. Incorporating a humidistat controller will allow fan speed to respond to a number of occupant-driven user conditions, which could attenuate short-term moisture burdens. A manual over-ride reversing airflow would also allow for summer cooling and ETS extraction during winter social 'functions'. As the thermally massive walls will maintain internal temperatures, increases in natural ventilation rates (opening windows to allow cross ventilation) can occur periodically when required. As the fan power requirements are relatively modest, excess electricity can be stored in simple car batteries to maintain ventilation rates during the night. In darkest winter the grid will be required, but the aim is to reduce demand and cost-effectively displace hydrocarbon and electricity usage.

Sustainability

The dwelling prototype (Figures 11.8–11.14 for plans, sections, elevations and perspectives) has a footprint of *c*. 150 m² (four-person, five-apartment detached villa) that incorporates four distinct strategies addressing issues of sustainability.

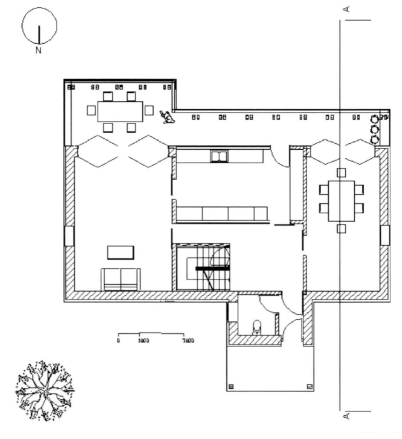

Figure 11.0 Ground-floor plan.

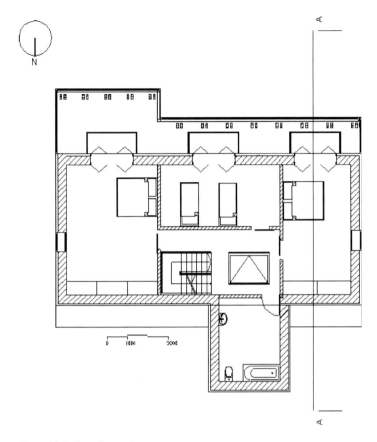

Figure 11 9 First-floor plan.

Reducing demand

Thermal capacitance combined with super-insulation negates the need for a central heating system. An NHER[6] audit of the prototype produced the maximum SAP and NHER rating of 100 and 10 respectively and a CO_2 burden of 1.9 t per annum. This equates to 0.475 t per person or $0.0126 \, t \, m^{-2}$. As the main space heating fuel source is timber and/or waste paper, the space heating demand will, in reality, have no net CO_2 burden. The programme, however, cannot address the additional benefits of back insulation. Modelling these as insulated opaque surfaces reduces the specific heat loss from 269 to 203 W/°C producing a heat loss parameter of $1.13 \, W \, m^{-2} \, °C$. Adding back the solar gains from the initial computer run (2293 W – utilisation factor of 0.82) results in the dwelling only requiring heating when the external temperature falls below 4.2 °C (i.e. the difference

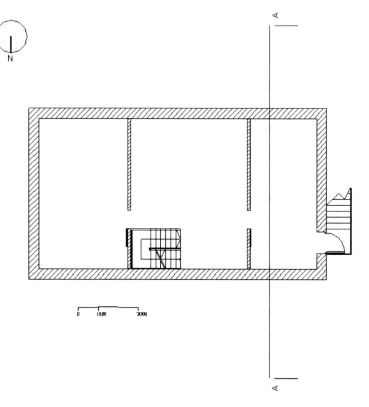

Figure 11.10 Basement plan.

between the demand and the solar gains, to a base temperature of 15.5 °C). The heating season is thus almost entirely reduced to winter nights in December–Febuary. The large thermal capacitance may, however, allow the solar utilisation factor to increase. The solar panels are predicted to displace about 45% of electricity or wood-fuel use for water heating. The multi-fuel back boiler can supplement water heating in winter and provide a central energy input for redistribution via the MHRV system to the hypocaust. Although a detached dwelling is illustrated, the construction techniques, features and systems are equally applicable to flatted developments which will have an even more favourable surface area to volume-enclosed ratio and a lower heat loss parameter.

Recycling, reuse and reversibility

The dwelling uses materials that can be reused or recycled and in the main will reverse to their original state (timber/claycrete/clay plaster/glass/steel). The design is also based on ensuring 'buildability' and 'demountability',

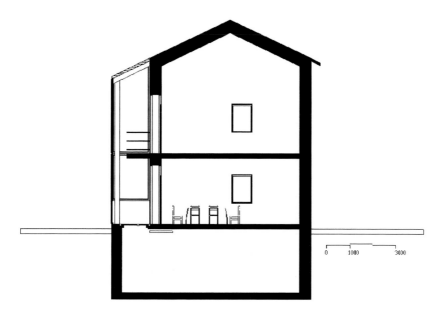

Figure 11.11 Section.

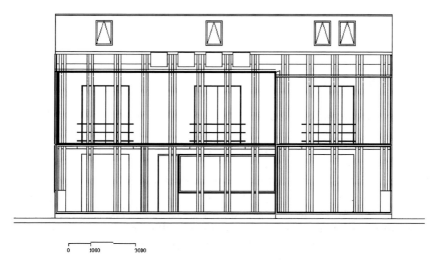

Figure 11.12 Front elevation.

with all components able to be joined and detached with ease (i.e. no wet trades or nails). The external cladding is a simple rain-screen mounted on horizontal rails on composite sheet bracing. A range of materials can be used to form the outer skin: timber cladding or recycled materials such as glass/slate/stone/metal/composite panels. The hung panel system would

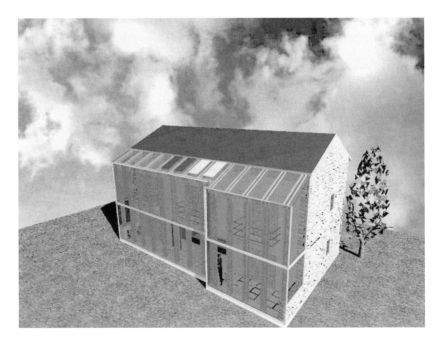

Figure 11.13 Perspective.

Figure 11.14 Sun buffer zone.

also allow periodic replacement and/or inter-changeability. The timber and steel frame has bolted connections and can be assembled and demolished with a spanner. Insulation can be fitted tightly in place using web flange rebates and traditional timber wedges. Clay plasters or clay boards should be used in all wet zones to help absorb moisture peaks – clay having remarkable hydroscopic properties that can suppress RH. Clayey loam (15-mm thick) can absorb close to 250 g of water vapour per square metre in 48 hr.[7] Clay plaster with coir can absorb $c.$ 75 g m^{-2} of water vapour in 48 hr, compared with $c.$ 45 g m^{-2} for gypsum plasters.

The surface finish is also important in determining the rate of moisture diffusion, even where the wall has the capacity to absorb when indoor relative humidity peaks. When compared with a 1 m thick layer of air (index value of 1) indoor emulsion paints have a value of 200–1500, and indoor solvent-based gloss paints 3000–5000.[8] When compared with concrete and brick, clay walls can absorb five to seven times the amount of water vapour. Using exposed clay plasters incorporating a colour pigment will thus allow the best result for both diffusion and absorption. Another benefit of this approach is that scuffs and accidental surface damage will be less obvious, as the pigment is spread equally throughout the plaster layer.

Replicability and prefabrication

To achieve economies of scale and quality control the dwelling has been designed to be pre-fabricated off-site and simply assembled on-site. The foundations are essentially a claycrete or limecrete basement raft-tank, that is lowered onto a layer of levelled and stabilised rigid insulation. The 'claycrete' walls extend to form the ground floor and north/east/west walls to first floor level. Some minor steel reinforcement will be required to lifting loads and to accommodate poor ground conditions. A steel frame is bolted onto this and provides support along the south elevation for the claycrete hypocaust floor panels. This structure is then encapsulated with a timber composite web column/rafter system, which incorporates 300 mm of wood-based semi-rigid insulation. The aim is to build the dwelling in less than four weeks with the main, sub- and super-structure taking less than five days to erect.

Avoiding HDM habitats

In general floor finishes, bedding and soft furnishings will provide the most suitable habitats for HDM colonisation. As previously discussed hard floor surfaces such as timber or linoleum (linoleum being the only material other than human skin which becomes less slippy when wet) provide less than ideal micro-climates and are easily cleaned. Where carpets are the chosen floor finish, under-floor heating coils should be incorporated which will reduce RH in the pile depth and inhibit colonisation. At a working temperature of $25\,°C$ the mixing ratio can rise to almost $15\,g\,kg^{-1}$ of dry air and still maintain the micro-climate below the HDM's critical equilibrium humidity.

Bedding should be encapsulated with micro-weave envelopes and duvets and pillows regularly washed at temperatures above $70\,°C$, which will denature any HDM allergen. Other simple design measures, such as extending all cupboards to ceiling level, reduces the area for dust shelves. Soft toys should be avoided or put out with the cat at night.

Knowledge in itself will of course not necessarily have a major impact on the volume builders who will continue to produce a poor quality, but highly profitable standard product. Some architects and clients have, however, been able to produce innovative designs with low CO_2 burdens. The BEDZED development in south London has incorporated some of the strategies and systems outlined above to good effect. Figure 11.15 demonstrates the integrated low/renewable energy thinking behind the concept.

It combines super-insulation (Figure 11.16) with high levels of thermal mass, a wind/stack-driven heat exchange ventilation system (Figure 11.17) and generates its own electricity on-site using a bio-mass generator.

Such developments have demonstrated that it is possible to produce zero-CO_2 designs and a quality environment, within existing housing association cost yardsticks (Figure 11.18).

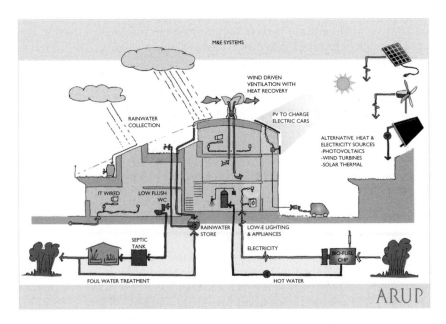

Figure 11.15 BEDZED energy systems and strategies.
Source: Bill Dunster Architects, Chris Twinn, courtesy of Arup

Figure 11.16 Super insulation and heavyweight internal blockwork.

Figure 11.17 Wind-driven heat exchange ventilation system.

Figure 11.18 BEDZED external view.

References

[1] Eurisol – UK Mineral Wool Ass. Reducing the Greenhouse Effect by Domestic Insulation. St Albans, p. 13.
[2] Natural Building Technologies, http://www.natural-building.co.uk.
[3] Porteous CDA. The potential for sun buffer zones in Scottish housing. *The International Journal of Ambient Energy*, 1985, Vol. 6, No. 3, pp. 137–146.
[4] Porteous CDA and Ho HM. Do sunspaces work in Scotland? Lessons learnt from a CEC solar energy demonstration project in Glasgow. *The International Journal of Ambient Energy*, 1997, Vol. 18, No. 1, pp. 23–35, ISBN 0143–0750.
[5] Sykes JM. Sick Building Syndrome: A Review, Health and Safety Executive (HSE), 1988, Bootle, Liverpool, p. 3.
[6] National Housing Energy Rating Scheme – National Energy Services Ltd, Rockingham Drive, Linford Wood, Milton Keynes.
[7] Minke G. Earth Construction Handbook, ISBN 1853128058, p. 17–28.
[8] Bablick and Federi. Fachwissen fur Maler und lackierer, Stamm Verlag, Koln, 1997 (see also www.igoeb.de).

Chapter 12

A fiscal strategy:
Fuel tax hypothecation

If the reader accepts the central thrust of this book – that dwellings in the UK are the most significant independent factor driving the current asthma pandemic – a fiscal strategy is required to be developed to underwrite retrofit interventions to the existing stock. This chapter will outline the main socio-economic generators which have made the problem more acute, quantify the possible scale of the current cost penalties being incurred and suggest a fiscal strategy which may go some way to ameliorating the problem within a realistic timescale.

The Scottish Energy Study

The Scottish Energy Study[1] published in 1993, investigating energy supply and use in Scotland, concluded that, 'there is a tendency for buildings to be designed to minimise capital costs, with little regard for lifetime running costs'.

'Short-termism' is of course, inherent when 'free market' economics is the accepted fiscal ideology and although it may well bring immediate benefits to small but powerful groups in society, this will be at the long-term expense and detriment of society as a whole. If capital expenditure decisions are based on pay-back periods of two or three years – a not uncommon scenario in the field of energy efficiency – it will reduce the public-sector borrowing requirement for capital works. The downside is that the year-on-year revenue account will increase. The so-called public–private partnerships (PPPs) are equally detrimental as they avoid capital spend, but future generations will have to pay the mortgage.

The causal links between housing and health are generally well established. The increasing cost of treating patients – presenting with essentially preventable conditions – will continue, unless the causes are addressed.

The chaotic 'free market'

In the mid-1970s, concurrent with, or precipitated by, the OPEC oil crisis, Harman[2] argued that the consensus around 'Keynesian' economic orthodoxy fell apart as recession bit deep. Into this vacuum stepped the 'new classicists' such as Friedman and von Hayeck, who rejected government intervention and preached a 'monetarist free market' doctrine, which reduced the role of the state to simply controlling the money supply and preventing the formation of the so-called unnatural monopolies. As the history of capitalism is one of merger and takeover to form larger monopolistic conglomerates, there does not appear to be anything unnatural about cartel formation. The primary driver is the accumulation of capital; monopoly is the end goal; Park Lane and Mayfair with four houses and a hotel. The stations and utilities, however, appear to be providing a more lucrative and secure return. People can choose to stay in hotels, but they have to drink water, and heat and light their homes.

During the 1980s the Thatcher government implemented a wholesale privatisation programme selling off Britain's energy, transport and telecommunications infrastructure, in an attempt to reduce public-sector spending in real terms. These new private companies were charged with producing a competitive return for their shareholders and thus their performance – and more importantly, the success of the overall strategy – will be judged by profit margins and dividends. Privatisation has simply involved auctioning monopoly franchises.

Taking the transport industry as an example, at a time when European business interests are implementing plans for closer integration and centralism, Britain has embarked on a course of deconstructing integrated networks and cross subsidies, with even through-ticketing proving to be problematic. At the end of 1998 the Central Rail Users Consultative Committee[3] reported that delays and cancellations were up by 22% and complaints by 103% on the previous year. Competition in the bus industry was also short-lived. In many cities large aggressive operators initially ran buses free on certain routes to undermine the competition and establish a monopoly. In 1995 Central Statistical Office[4] reported that the number of passenger journeys had reduced and the cost per mile had increased in real terms. In November 2003 the author was unable to get a return fare from Glasgow to Leicester – two of Britains major cities – as three different train companies were involved. The eventual cost was three times that of a quoted return airfare to London.

The political framework

Callinicos and Simons[5] have argued that such wholesale privatisation would not have been possible without the implementation of the *Ridley* plan which was concerned with the systematic destruction of the National Union of Mineworkers. The document, which was leaked to *The Economist*,[6] involved strategic action on several fronts – the rigging of capital figures to make coal look an uneconomic option; contingency plans for importing coal; stock-piling at power stations; the recruitment of non-union drivers and the formation of a large integrated 'national' police force equipped and prepared to neutralise picketing. It also involved the expansion of the nuclear power industry. Figures were presented which purported to show that nuclear power was the cheapest option. On any proper comparison[7] taking into account the capital debt of building new nuclear capacity, nuclear-generated electricity in the mid-1980s cost 3.2 p per kWh against 1.8 p for coal. Furthermore, the continuing problems of the now renamed British Energy, has required a massive injection of public money, first to pay off the majority of its capital debt burden, then to bail it out when it declared that its operations were not commercially viable. It is unable to compete in the market place. When it comes to energy supply and distribution, the playing field does not appear to be level.

The opening-up of the domestic fuel market has led in the short term to gas and electricity becoming cheaper, as the various suppliers scramble for market share. Making fuel cheaper further delays capital investment in the 'clean' and/or renewable technologies and power generation. As such, privatisation is clearly an obstacle to integrated and rational long-term central planning, especially at a time when the First World is eventually starting to, at least, discuss global warming scenarios.

The Earth Summits

The current international debate on global warming reflects the inherent conflict between commercial vested interest and precautionary rational planning. The Rio, Kyoto and Johannesburg summits failed to produce binding international agreements. In Rio, deadlock was reached on every key item after 15 solid weeks of preparatory negotiations. As the summit itself unfolded, a clear North–South polarisation was increasingly expressed through two counterposed arguments over the responsibility for the crisis. On the one hand, the North's over consumptive industrial 'model' of development was attacked by the 'third' world countries, while the South's inability to manage its population explosion was attacked by Europe and the US. While the scientific community called for an immediate 60% reduction in atmospheric emissions, the most that was agreed was an open-ended, non-legally binding statement of 'intent' to hold emissions at 1990 levels. The US whose industries are responsible for 25% of global pollution, with Britain's open support, managed to delete the specific target date of 2000 from the treaty, previously agreed by 110 countries.[8] The Johannesburg summit descended into utter fiasco with no agreement being signed on any substantive measures.

Energy efficiency investment criteria

The hottest year since records began was 2003. Global warming is predicted to result in rising sea levels due to the Antarctic ice-cap melting, combined with general thermo-fluid expansion. The jury remains out on the rate at which this will occur. Clayton[9] has estimated that a rise of 1 m would affect some 5×10^6 km^2 of land surface containing 30% of the world's current productive cropland. Despite the size of the threat, all of Britain's main political parties believe that market economics will have to generate the solution. So what criteria are commerce and industry using to address current energy consumption? The National Westminster Bank was one of the first companies to sign the Energy Efficiency Office's 'Making a Corporate Commitment' campaign. In a publication[10] to highlight its achievements on effective energy management, it states that 'candidate projects for capital investment will only be considered on the basis of a three year or less pay-back period'. Its energy policy recognises that, 'energy use is secondary to the business objectives of the bank and is phrased to ensure that energy efficiency does not impact negatively on business performance . . .'.

In 1993 they invested 1.235% of their energy bill (the energy bill represents less than 1% of the bank's turnover) in capital measures. No objective performance indicators were published in the report. If the same criteria were to be adopted in the domestic sector, based on the 1996 Family Expenditure Survey,[11] it would allow the average householder to invest no more than 83 p per annum. An investment of such modest proportions hardly justifies the printing of glossy brochures, trumpeting 'green' credentials. Like so much of the current hype surrounding 'corporate sustainability', this is no more than a crude marketing exercise, and should not be taken seriously.

The real cost of fuel

In 1992 the Department of Trade and Industry[12] published figures demonstrating that the additional external costs of fuel cycles for electrical power generation, were between 0.5 p per kWh for new gas-fired power stations to 5.75 p per kWh for old oil-fired power stations. Although the marginal stations are only brought on stream to cope with peak demand, and the situation is further confused with nuclear power stations providing the base load, there is a coherent and cogent justification that if – as the government claims – the 'polluter has to pay', these environmental costs should be included in the market price. Although industry and commerce will lobby against what will be perceived as a carbon or environmental tax, in reality, it is no such thing. It is simply the intervention of yet another regulator to ensure that the real and calculable external costs are included in the price. Energy can thus be considered to be unrealistically inexpensive and when combined with 'free market' economics, results in short-term decision-making. Burning gas in power stations is yet another short-term palliative that results in the loss of almost 50% of its calorific value and further quickens the depletion of an exceptionally valuable 'clean' fuel.

Fuel poverty

There is however, a major problem with adopting an approach which simply adjusts the pricing mechanism. Satsangi, Malcolm and MacIennan[13] showed that the poorest decile of Glasgow's population spend, on average, a remarkable 24% of their net income on fuel (this compares with 3.2% for the richest decile). A report by Energy Action Scotland[14] into the effects of VAT on fuel poverty confirmed the fuel poor's predicament; they are more likely to be living in poorly constructed, uninsulated dwellings without central heating and a high proportion are forced to use 'expensive' domestic tariff electricity for heating purposes. Between 1991 and 2002, although the prevalence of condensation has reduced from 19.3 to 11%,[15] the tendency for the poorest section of the community (income of less than £100/week) to live in the least energy-efficient dwellings remains. Furthermore, the 2002 Scottish House Condition Survey reported that despite some improvements in energy efficiency combined with a significant drop in real fuel costs, 369 000 (17%) households remain in 'fuel poverty'. As the initial scramble for market share by the privatised utilities is now coming to an end and a carbon tax may not be far off, most economists are predicting fuel price inflation. The next few years may very well see those on the margins being re-classified as 'fuel poor'.

The most comprehensive of social surveys have consistently shown that rather than wealth and prosperity trickling down, poverty has been surging up. In 1996 the Office for National Statistics published figures showing that since 1971 the percentage of the population surviving on less than half the average income levels increased from 11 to 21%. In 2003 the Scottish Executive[16] published a new poverty assessment entitled, the Scottish Index of Multiple Deprivation (SIMD). It attempted to measure and index by region, five key parameters: income, employment, health and disability, education, skills and training and geographical access to services. Figure 12.1 presents the index by rank order, with Glasgow having an SIMD score of 46.88 and East Dunbartonshire, the least deprived region, with a score of 9.07.

During the period 1997 to 2002, Scotland registered 16 660 excess winter deaths (EWD is taken as the additional deaths during December–March than occurred in the preceding and subsequent four-month periods) in the population over 65. The number of deaths attributed to lung cancer over the same period was 23 938. In terms of impact on mortality EWD's are not quite as influential as smoking, but may have a similar influence on morbidity,

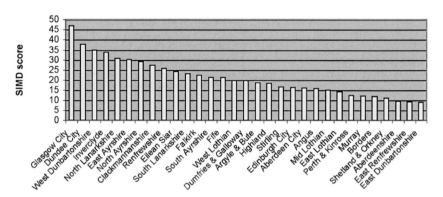

Figure 12.1 SIMD by region.

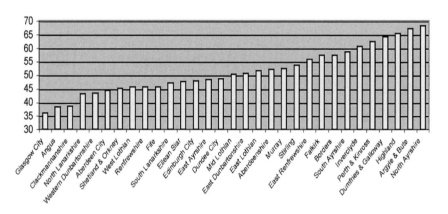

Figure 12.2 Odds of becoming an EWD by region.

as cold stress reduces the immune system's ability to fight infections. When broken down by Scottish region (Figure 12.2) the chance of becoming an EWD in Glasgow is 1:36 rising to 1:68 for East Dunbartonshire, at the other end of the scale.

The SIMD is positively correlated with EWD by region (0.35 at the 5% confidence level). Although diet and lifestyle are important variables, this correlation appears to go against the influence of climatic variations, domestic energy efficiency and access to the gas network.

Climatic variations

There is a significant temperature gradient across Scotland with Dumfries and Galloway in the south west having the mildest climate (c. 2 300 degree days – a degree day is a measure of how often the external temperature drops below 15.5 °C – considered to be the base temperature at which domestic heating is required) and Shetland in the north east, having the coldest temperatures with higher wind speeds (over 2 900 degree days). Mean 30-year average daily temperatures for January show a variation of 6 °C across Scotland, with temperatures dropping as you go north and away from the coasts. As EWD increase linearly with temperature fall, these colder regions should have significantly higher winter death rates, if outdoor exposure was the key causal factor. Although Figure 12.2 clearly shows that the death rates along the south western seaboard (Dumfries and Galloway, Ayrshire and Argyle and Bute) are at the lower end, this does not appear to hold good for the rest of Scotland, as Glasgow and Dundee fall within relatively mild climate zones.

House type and home energy ratings

The size, configuration, construction type and the efficiency of the heating provision will be influential in the occupant's ability to maintain healthy indoor temperatures. Where indoor temperatures regularly fall below 16 °C, 'thermal safety'[17] will be compromised. Those living in large, uninsulated dwellings with expensive and/or inefficient heating systems will be penalised. The configuration and external surface area will also be influential. Urban locations have significantly more flatted dwellings. A mid-floor flat that has an excellent ratio of surface area to volume enclosed, will typically only lose heat through small areas of external wall/window. A detached rural property could have four times the heat loss parameter (W/m^2°C floor area) as heat is being lost through the floor and roof, as well as four external walls. The EWD by region bucks this trend with the major cities (in particular, the dense tenemental cities of Glasgow and Dundee) being at the top end of the scale.

Cost of fuel

Similar to energy efficiency, those who have access to relatively cheap gas can purchase affordable warmth. Even when combusted in old or inefficient appliances, gas will still provide a kW of heat for under 2 p (closer to 1 p in condensing boilers); compared with c. 3.5 p/kW for off-peak electricity and over 7 p/kW for the domestic on-peak tariff. The 2002 Scottish House Condition Survey[18] estimated that over a quarter of dwellings (612 000) do not have access to mains gas. Rural, northern and island communities are thus penalised, but the statistics by region do not have the sensitivity to isolate this as a key variable.

Diet and lifestyle

A diet high in fat and salt will play an important role in determining heart disease, stroke and possibly even allergy. Alcohol abuse, smoking rates and drug-taking are not restricted to urban locations. When the total population of a region is divided by the deaths rates from chronic liver disease and cirrhosis, Glasgow has a rate of 1 in 441 (1310 deaths from a total population of 577 869). This compares with the lowest incidence in the Western Isles of 1 in 2 944 and the highest rate in Highland at 1 in 122. The picture differs for lung cancer death rates (malignant neoplasm of trachea, bronchus and lung) however, Glasgow has a rate of 1 in 143, which compares with 1 in 182 for Dundee, 1 in 248 for more affluent East Renfrewshire and 1 in 523 for the Western Isles. Although external air quality may play a role in lung cancer deaths it appears that smoking is still more prevalent in the urban working class. There are no regional statistics for fish-supper consumption rates.

Housing and health

Work by Hunt *et al.*[19] published in 1989 established a causal link between dampness and a variety of medical complaints, particularly those associated with the respiratory tract, while Boardman[20] uncovered an underlying relationship between winter temperatures and excess winter deaths and health morbidity. A distinct and significant causal relationship was thus established between housing conditions and other external costs, specifically in the form of increased hospital admissions (ischaemic heart disease, stroke and respiratory disorders), prescription charges, medical consultations and absenteeism.

Since 1991 – the year of the first report into Scottish housing – gross public housing investment in Scotland remained relatively stable up to 1996/97 when it dropped from a maximum figure of £1228 million[21] to a predicted low of £761 million.[22] Yet after approximately £5 billion being spent on the stock in five years, this level of investment failed to make any real impact on the dampness statistics, with 534 000 dwellings still affected by dampness and/or condensation. In 1996 the Scottish House Condition Survey[21] concluded, that there was, 'no substantive change'. It would thus appear that a significantly higher level of investment is required.

Introducing a domestic fuel tax

Introducing a household fuel tax incorporating a tax-free threshold to be set at around £600 (based on a retrospective analysis of the previous 12 months' fuel bills and collected by the supply companies) would have little or no effect on the lower five deciles of the current household expenditure profiles (Office for National Statistics 1996). This equates to an average weekly expenditure of £11.53 on tax-free fuel (gas £4.50, electricity £6.20, other £0.83 NB 1995–1996 prices are similar to 2002–2003 cost indices) before a base rate of 50% is applied. The total additional tax burden to the domestic sector ranges from £6.50 per annum for decile four, to £130 per annum for decile ten. If a further threshold is introduced at say, £1000 per annum to be taxed at 100%, profligate energy users will be penalised and payback times for capital measures should shorten considerably. The above taxation regime should raise a figure approaching £1 billion per annum, which should be earmarked and returned to the domestic sector as grants to support public and private home insulation rolling programmes, such as the Home Energy Efficiency Scheme (HEES).

The industrial and commercial sector could also be subject to a similar structure with the tax-free thresholds geared to ensure that companies are rewarded for significant increases in efficiency or reductions in consumption. At least half of the tax yield could be returned to this sector in the form of targeted incentive grants to encourage a sea change in corporate behaviour, while the remainder is ploughed into wider environmental clean-up projects or the research and development of clean and/or renewable energy projects. If the tax can be hypothecated, the increased yield can be returned to the sector from whence it came, in the form of capital investment grants to fund a cost-effective, ten-year rolling programme to insulate and/ or refit. As the tax yield should reduce proportionally over time with improved energy efficiency, a simple yet attractive exit strategy is incorporated in the proposal.

Insulating the existing housing stock

In 1989 Eurisol[23] estimated that it would cost under £35 billion to retrofit the existing British housing stock generating £2.334 billion in energy savings per year (pay-back period 14.9 years). Increasing the cost of fuel will clearly reduce the pay-back period. Insulating the domestic sector alone will reduce CO_2 emissions from over 500 million tonnes, to under 200 million tonnes – the reduction of over 60% being called for by the world's scientific community. A cost-benefit strategy would take into account the reduction in power station output. Although electricity only contributes 5.3% of total space heating load, at least two power stations could be taken off-stream or would not require to be replaced. When combined with appropriate ventilation regimes and techniques, such as heat reclamation fans, the effects on public health and a reduced burden on the health service could have a realistic and manageable pay-back period. In effect, such a strategy could more than double our fuel reserves and buy vital time to develop and implement a range of renewable energy options that will allow future generations to live useful and meaningful lives.

The future of housing and the bigger picture

Between 1991 and 2001, private house builders in Scotland, completed between 14 100 and 19 000 dwellings per annum. Almost all these dwellings were of timber frame construction, incorporating polythene vapour barriers. Housing Associations (HAs) completed c. 2100 dwellings in 1991, rising to almost 6000 by 2001. Since 1995, Local Authorities (LAs) have built, in total, less than 500 new dwellings.[24] In 2001 there were reported to be 2.345 million dwellings in Scotland.[25] Private housing represented 70.9% of the stock, HAs 6.3% and LAs 22.8%. Right-to-buy legislation and recent stock transfers (the City of Glasgow transferred 81 000 dwellings to the Glasgow Housing Association on 7 March 2003) have resulted in further erosion of the publicly owned stock.

The economics of housing have thus, come full circle in the twentieth century. The market has reasserted its control over housing provision, but at the centre of the debate around the so-called sustainable development is a clear obligation on current generations not to enrich themselves at the expense of future generations. Free-market economic models appear to be structurally incapable of reforming or adjusting the modes of production, distribution and exchange by the required scale. Indeed, it is already apparent that 'business interests' have lobbied hard at local, national and international level to sabotage almost all rational planning initiatives in the interests of short-term profit. 'Short termism' is built into the market system and will continue to prevent more rational investment strategies and planning. After declaring initial high profit margins, Britain's nuclear generator has required massive state intervention to stop the company from going into liquidation. The state, which used to subsidise housing to achieve reasonable standards, is now subsidising private companies.

Without political intervention, the market will continue to produce economic chaos and deleterious effects on our macro-, meso- and micro-climates. The price being paid by the growing legions of asthmatics may only amount to a small proportion of the total bill. The global catastrophe that destroyed Honduras in 1999 and the floods that devastated central Europe in August 2002, may be more indicative of the price future generations will have to pay for the current complacency over energy production and use.

Are landlords liable?

There has been a considerable number of cases where tenant's have taken action against their landlords for damages – both to health and property – due to suffering cold and damp housing conditions. These cases have normally been founded upon the failure of the landlord to provide a dwelling that is 'in a tenantable and habitable' condition, fit for human habitation as contained in the Housing (Scotland) Act 1987. In terms of Schedule 10 of the said Act,

> there shall be implied a provision that the house will be tendered to the Pursuer in a tenantable and habitable condition at the commencement of the tenancy and an undertaking that the house will be kept by the landlord during the tenancy in all respects reasonably fit for human habitation.

The crave would normally include a sum identified in the pleas in law, to recompense the pursuer for suffering loss, injury and damage. Injury of course covers the area of health, and where the burden of proof supports the view that a tenant has developed asthma (or any other disease such as mesothelioma caused by asbestos fibres) due to exposure to HDM proteins, there appears to be clear legal precedent. As there is a proven dose–response relationship between HDM allergen levels and the development of asthma, where a dwelling is found to have reservoirs above the known thresholds, and the tenant has a demonstrable bronchial reaction (skin prick or challenge tests), causality is demonstrable. Compensation to date, for tenants winning such actions has been relatively modest with pursuers being awarded a few thousand pounds in damages for ill-health. The courts may not take such a miserly view where it can be proven – on the balance of probability – that living in a particular dwelling has led directly to the development of a life-threatening and highly debilitating disease; a disease that will affect all aspects of an individual's quality of life from birth to death.

The Occupiers' Liability (Scotland) Act 1960, also covers a tenant's children who may be most at risk. In the case of *Scott & Scott* v. *Glasgow District Council*[26] the relationship between damp living conditions and child asthma was put to a 25-day proof at Glasgow Sheriff Court. Sheriff Wilkinson found that child asthma was materially exacerbated or triggered by damp living conditions. The case hit the legal buffers because the pursuer had set her stall on a 'common law' reparation claim, while making a brief mention of the council's 'statutory duty'. The court held there was a complete failure to specify a statutory case. This 'winning on the facts' but losing on the pleadings was made all the more unfortunate when the pursuer's minute of amendment was refused after the evidence. Among other things, the case of Scott illustrates very powerfully the need to plead a proper Occupiers' Liability (Scotland) Act claim, in non-tenant personal injury cases.

As many insurance companies have recently woken up to the financial risks associated with poor indoor environments and excluded liability for the deleterious effects on health that mould growth can produce, landlords may find themselves having to pay such damages from the revenue account. What price will a court place on a lifetime struggling for breath?

References

[1] AHS Emstar. Scottish Energy Study: towards an energy policy, Glasgow, Autumn 1993.

[2] Harman C. The Crisis in Bourgeois Economics. *International Socialism*, Issue 71, London, 1996, p. 5.

[3] Central Rail Users Consultative Committee, 26 August 1998, London.

[4] Central Statistical Office, Social Trends 1995 edition, HMSO, London, pp. 201–214.

[5] Callinicos A and Simons M. The Great Strike, Bookmarks, London, 1986, p. 36.

[6] *The Economist*, 27 May 1978.

[7] Callinicos A and Simons M. The Great Strike, Bookmarks, London, 1986, p. 40.

[8] Treece D. Why the Earth Summit failed. *International Socialism*, Issue 56, London, 1992, p. 65.

[9] Clayton A. Global Environmental Change: The Challenge for Planners and Architects. Institute of Policy Analysis and Development. Edinburgh University Press, 1993, p. 40.

[10] Energy Efficiency Office. Department of the Environment, Case Study 259 – Energy Management at National Westminster Bank plc. HMSO, December 1994.

[11] Family Expenditure Survey, Office for National Statistics, Family Spending, HMSO, 1995–96, Section 1.3, Detailed household expenditure by gross income decile group.

[12] Department of Trade and Industry. Social cost of fuel cycles – UK Environment externality adders for electrical power generation, HMSO, London, 1992.

[13] Satsangi M, Malcolm J and MacIennan D. Glasgow House Condition Survey, Staying Dry, Keeping Warm and Dampness. Centre for Housing Research, University of Glasgow, 1991.

[14] Energy Action Scotland. Fuel Poverty in Scotland, Glasgow, 1995, pp. 13–19.

[15] Scottish Homes/Communities Scotland. Scottish House Condition Surveys 1991 and 2002, 1992/2003, Edinburgh, ISBN 1 874170 54 1.

[16] Scottish Executive. Scottish Index of Multiple Deprivation, Edinburgh, 2003.

[17] Howieson S. Housing – Raising the Scottish Standard, Technical Services Agency Ltd, 1991, Glasgow.

[18] Communities Scotland. Scottish House Condition Survey 2002, Edinburgh 2003, ISBN 1 874170 54 1, Chapter 1, p. 10.

[19] Hunt SM, Martin CJ, Platt SD, Lewis C and Morris G. Damp housing, mould growth and symptomatic health status. *British Medical Journal*, London 1989, Vol. 298, pp. 1673–1678.

[20] Boardman B. Fuel poverty: from cold homes to affordable warmth, Belhaven Press, London, 1991.

[21] Scottish Homes. Scottish House Condition Survey 1996, Edinburgh, 1997.

[22] Wilcox S. Housing Finance Review 1997/98, Joseph Rowntree Foundation, 1997.

[23] Eurisol – UK Mineral Wool Ass. Reducing the Greenhouse Effect by Domestic Insulation. St Albans, p. 13.
[24] Scottish Executive Statistical Bulletin – Housing Series, HSG/2003/1, Edinburgh, February 2003.
[25] Energy Action Scotland. *Parliamentary Bulletin*, Issue 55, Scottish Executive News Online, 7 March 2003.
[26] *Housing Law Review*, 1997 (d18.1, 21.6, d30.6).

Chapter 13

Conclusions and recommendations

The nature of the research exercises and literature review undertaken and reported in this work can be composited into four primary and six secondary research questions and conclusions.

What historical changes and key drivers have led to the HDM species becoming so common in our domestic environment?

Chapters 4 and 5 reviewed the most significant historical drivers and changes that have influenced the design and production of the UK housing stock during the twentieth century. Key changes in construction techniques and built form were modelled and the likely effect on air change rates and volumetric airflow was estimated. The comparative reduction in both modelled and measured ventilation rates was striking and – when taken in conjunction with the experimental work undertaken by the BRE reviewed in Chapter 10 – lends support to this as a clear and measurable trend, likely to have produced more humid domestic environments.

In combination with changes in occupant behaviour and lifestyle – such as the almost ubiquitous use of wall-to-wall carpeting – the increase in internal water vapour pressures is likely to have resulted in the provision of suitable habitats and micro-climates for HDM colonisation and proliferation. The research findings reported in Chapter 7, suggest that many dwellings in west central Scotland contain high levels of dust mites and their associated allergens. Such a hypothesis is reinforced by the literature reviewed in Chapters 1 to 3, which confirmed that the allergenic proteins contained in HDM faecal pellets, can regularly be found at concentrations in the dust reservoirs known to cause asthmatic sensitisation and trigger symptoms. If the levels found in the dust reservoirs of the dwellings tested in the research programme are typical of the Scottish and/or UK housing stock, a significant part of the current asthma pandemic (80%) is highly likely to be primarily driven by house design, construction and use.

Recommendation

Building standards and codes should be revised to prioritise moisture control as being key to health and safety. A performance specification can be framed that requires dwellings to be capable of maintaining – given normal domestic activity regimes – internal RH below 63%; the critical equilibrium threshold of ambient air above the carpet, required for HDM colonisation and proliferation. The use of variable-speed MHRV units controlled by humidistats in small, 'tight' dwellings, appears to be a technical fix which will – in most cases – overcome the deficiencies of the design and could be recognised as a 'deemed to satisfy' approach.

What can be done to reduce allergen levels in the existing and future domestic environment?

The results of the initial research programme confirmed that allergen-avoidance regimes and high-pressure steam cleaning can encapsulate and denature a significant proportion of the allergen burden (c. 92–94%) in the dust reservoirs. MHRV systems can also reduce internal water vapour pressures. As the RH in carpets is likely to be 10% higher than ambient air, maintaining RH below 60% will ensure that the HDM's critical equilibrium humidity will not be achieved in the micro-climates.

Any significant rise in RH – such as that caused by diurnal temperature fluctuations in lightweight dwellings – is highly likely to facilitate colony viability and stimulate increased metabolic activity and allergen production. Although a mixing ratio below $7\,g\,kg^{-1}$ of dry air will inhibit HDM activity, maintaining internal temperature stability between the 18 to 21 °C comfort recommendation will – given normal moisture production regimes – allow for a degree of flexibility (60% RH at 21 °C will allow the mixing ratio to rise to c. $9.4\,g\,kg^{-1}$ – at 18 °C a mixing ratio of $8\,g\,kg^{-1}$ of dry air).

Recommendation

As construction techniques are producing much tighter dwellings, 'fortuitous' background air infiltration or leakage rates will be significantly reduced. Where dwellings are relatively 'small' and 'tight', mechanical ventilation – incorporating cross-flow heat recovery devices – will allow ventilation and associated vapour diffusion rates to be increased to a rate that suppresses the mixing ratio below $7\,g\,kg^{-1}$ of dry air during the winter months, without incurring a severe energy-cost penalty. The planned ventilation strategy should represent, as closely as possible , the actual air change rate. Designing dwellings with a high level of insulated thermal mass – while optimising solar and casual gains – can provide a stable thermal environment and reduce diurnal temperature fluctuations, presently implicated in overnight micro-climate moisture gain.

What other factors can influence the efficacy of the remediation and measured outcomes?

Measuring the health impact of such an interventionist trial has to take account of a plethora of confounding variables. Although allergen avoidance and denaturing techniques can significantly reduce the level of allergenic proteins – both airborne and in the dust reservoirs – increasing ventilation rates may also reduce indoor concentrations of other predisposers, sensitisers and triggers. Any impact on respiratory health may thus be, in part, due to reducing or diluting indoor gas concentrations, microbes, bioaerosols, ETS, endotoxins and PM, and not simply the outcome of reducing allergen levels alone. It is highly likely that many dwellings are now under-ventilated and that the 'accepted' rate of $0.5\,\mathrm{ach}^{-1}$ is not adequate for vapour or pollutant dispersion or dilution.

Recommendation

Any research programme with aspirations to quantify the causal mechanisms must adopt a multi-factorial and multi-disciplinary approach and at least attempt to scope the main confounding variables. The development of a research base investigating the impact of indoor air quality on health, with a focus on the additive, synergistic and antagonistic effect of a range of key pollutants, is urgently required to form the basis for future indoor air quality guidelines. As many pollutants are known to be triggers for asthmatic incidents, such guidelines require to take into account short-term peak exposures, as well as longitudinal mean levels.

What is the scale of the likely costs and benefits of such a preventative approach?

The increase in health service costs directly associated with primary and acute treatment of asthma now accounts for close to £1 billion of the UK's health budget. The National Asthma Campaign claims that this figure can be doubled when costs for lost work and school days are factored into the equation. Adopting a preventative strategy that addresses both an environmental sensitiser and trigger, could result in savings in treatment costs to the current asthmatic legions. Ensuring that future generations have a reduced exposure to HDM proteins is likely to reduce the number of new cases. Medical records at present, do not appear to allow accurate data on patient cohorts to be systematically extracted. Individual case studies will require a much higher level of clinical monitoring as well as a linked drug 'step down' regime to produce a reliable and accurate measurement of costs and benefits.

Recommendation

Medical records and prescription details should be standardised and stored on a central database, sorted by disease category. Health Boards should give consideration to funding remediation programmes in asthmatic's dwellings. Landlords should be required to audit and undertake risk assessments in their properties for HDM infestation. Grants should be made available by local authorities for denaturing regimes and appropriate retrofit remediation.

Secondary outcomes

Targeting the domestic environment

The greatest exposure to dust mite proteins is likely to occur in the home. The literature suggests that both the workplace and external environment do not have either the micro-climates or climate to support large colonies. Any attempt to reduce exposure to HDM proteins should concentrate on the remediation of the domestic environment.

Adopting a new HDM allergen sampling protocol

The research exercise reported in Chapter 7, confirmed that HDM allergens are not equally distributed throughout the dust reservoirs. Subsequent testing, reported in Chapter 8, confirmed a 70-fold variation across a carpet in one room. Although the factors driving such asymmetry are at best contentious and at worst unknown, any interventionist trial which attempts to measure longitudinal changes must adopt robust sampling techniques that can measure the distributed mean level of *Der p1* allergen burden in a room.

To account for significant variations in background domestic dust levels, absolute amounts per square metre should be presented alongside ratios. To produce an average of the total *Der p1* burden in any given room, the entire exposed carpet area should be vacuumed at a minimum rate of $1\,\mathrm{min\,m^{-2}}$. Furniture and beds should also be vacuumed at the same rate. Research into the factors underlying such concentrations should aim to identify the main drivers. This is likely to require the more controlled environment of an entomology laboratory.

Clinical assessments

Although self-reported peak-flow data in combination with health questionnaires, can provide an evidence base to evaluate general health impacts, a much more detailed level of clinical testing is required to establish biological mechanisms and causality. Clinical testing should include skin prick and blood tests and reverse spirometry (FEV) in addition to twice-daily peak-flow readings.

Cost benefits

Assessing the cost effectiveness of such an intervention requires drug use to be both accurately measured and controlled. Any improvements in lung function can then trigger a step-down regime in inhaled steroid and $Beta_2$-agonist treatment and reduce the burden on both primary and acute health care systems.

Medical professionals will be required to take charge of patients' drug treatment regimes and should more accurately record the use of both primary and acute care facilities associated with the treatment of asthma and monitor lost workdays/schooldays.

Cohort selection

Young children are not ideal candidates to include in such studies for three reasons: the diagnosis of asthma may be provisional; varying growth rates will produce a natural increase in lung function and more detailed tests which require invasive techniques may produce higher attrition rates or non-compliance. Selecting an older cohort (>16) whose asthma is more predictable and relatively stable, can reduce this 'background noise' and allow any changes brought about by the interventions to be more accurately evaluated.

Any cohort should be carefully selected to ensure that dust mite sensitive and relatively stable, diagnosed asthmatics are encouraged to participate. Improvements in lung function and their association with specific interventions can then be made with a higher degree of confidence.

Airborne allergens and floor coverings

The relationship between floor coverings and indoor air quality remains relatively under-researched. It is vital that future research quantifies the particle size, absolute amount and the duration of suspended allergenic PM and investigates the relationship between the levels contained in the dust reservoirs and their suspension duration.

The literature reviewed in Chapter 8 and 9, supports the view that it would be prudent for those with asthma, or a congenital predisposition to allergy (atopy), to remove their carpets and live in a dwelling with easily cleaned, hard floor surfaces that do not off-gas volatile organic compounds.

Summary of key contributions

This book has attempted to link a variety of subject areas: housing legislation and economics; technical standards, construction techniques and energy efficiency; ventilation rates, indoor air quality, psychrometrics, micro-climates and entomology; immunology and respiratory medicine; fiscal strategies to underwrite cost-effective remediation techniques and the development of a new housing prototype designed to maintain healthy air quality and lower RH.

It has attempted to demonstrated that the historical tension between the implementation of minimum statutory standards and *laissez-faire* market forces, remains a contemporary issue. The pressure on space standards, combined with the drive for energy efficiency, has resulted in a significant reduction in internal ventilation rates. This in turn is highly likely to have led to a significant deterioration in indoor air quality and an increase in internal water vapour burdens. The generic house-type modelling and blow-door testing confirmed the likely scale of this reduction in dwellings common to west central Scotland.

The research programme tested a methodology for both quantifying the historical level of HDM activity and the efficacy of a set of remedial measures designed to reduce exposure to HDM allergens. Although parts of the study were statistically under-powered, the methods appeared to be relatively robust. Despite the plethora of confounding variables, the results supported the primary hypothesis – modern dwellings and living patterns are producing micro-climates suitable for HDM colonisation and contain a significant level of allergenic protein in the dust reservoirs. The allergen avoidance and denaturing intervention was able to reduce exposure to these proteins and generated statistically significant, if self-reported, health benefits. The trial also identified weaknesses with existing dust sampling protocols, which led to the development of a more robust multi-factorial protocol.

The literature review on air quality and air tightness confirmed that although the existing pre-1980s stock has ventilation characteristics that render them relatively 'leaky' and 'draughty', volatile organic compounds and other gases have been found in concentrations likely to produce deleterious effects on occupant health. Such a finding clearly undermines the existing 'agreement' that a ventilation rate of $0.5\,\text{ach}^{-1}$ will ensure reasonable air quality throughout a dwelling – or more importantly – across an individual room. The trend to produce even tighter dwellings – implicit in current Building Regulation revisions – will further exacerbate these trends.

Moisture control must now be considered as the crucial variable determining ventilation rates. The previous threshold of 70% RH (based on limiting mould growth) should be reduced to 60%, if HDM colonisation and proliferation in the micro-climates are to be avoided. It has also produced evidence to support the key role MHRV can play in increasing air change rates and suppressing humidity, without incurring an energy-cost penalty.

Indoor air quality is clearly an area that requires a much larger research focus. Increasing the evidence base could identify the most significant variables impacting on respiratory health. Appropriate and standardised measurement techniques require to be established, as will health risks from pollutants, both in isolation and in combination. Indoor air quality guidelines should be developed, agreed and enshrined in legislation.

The UK also requires to develop a fiscal strategy aimed at delivering increased funding to ameliorate problems caused by poor indoor air quality. A hypothecated fuel tax would be relatively simple to collect and has a politically attractive exit strategy, based on the virtuous cycle of simultaneously improving both domestic energy efficiency and air quality. Although remediation is possible, the exploration of design options from first principles, led to the development of a new prototype dwelling, designed to ensure 'healthy' air change rates, while suppressing humidity and dust mite activity. The design also addresses the associated concerns of energy use, pollution and collateral damage to the environment.

The book represents a comprehensive, if somewhat generalist, approach to housing and health. The series of research exercises completed and detailed, when taken in conjunction with the growing body of scientific literature reviewed, supports the view that – at least on the balance of probability – changes to the design and use patterns of our dwellings are highly likely to be the most significant single driver producing the current asthma pandemic in Britain. Future research requires a focus on primary preventative strategies, particularly as it appears likely that HDM allergens can cross the placenta and sensitise embryos before birth. There is clearly a range of complex factors that are predisposing a growing number of individuals to develop allergies. These factors will remain the remit of the medical professions. Indoor air quality and the suppression of HDM colonies fall within the remit of architects and engineers. It is their job to produce environments that do not have such negative impacts on public health. Eradicating HDM allergens and designing warm dry homes is not a technically difficult task. The research and knowledge base is now in place. There is no excuse for inaction. To do nothing is to condemn a growing proportion of our children to a life struggling for breath.

In his seminal work, 'Medical Nemesis – The Expropriation of Health' published in 1975, Ivan Illich was unequivocal about the limits of medicine and the need for a broader approach to public health,

> The professional practice of physicians cannot be credited with the elimination of old forms of mortality or mobidity, nor should it be blamed for the increased expectancy of life suffering new diseases. For more than a century, analysis of disease trends has shown that the environment is the primary determinant of the state of general health of any population. Medical geography, the history of diseases, medical anthropology and the social history of attitudes towards illness have shown that food, water and **air**, in correlation with the level of socio-political equality

and cultural mechanisms that make it possible to keep the population stable, play the decisive role in determining how healthy grown-ups feel and at what age adults die... One third of humanity survives on a level of under-nourishment which would formerly have been lethal, while more and more rich people absorb ever greater amounts of poisons and mutagens...

Index